Electronics and Computer Acronyms

To
Sue, Camilla and Estella
with thanks

Electronics and Computer Acronyms

Phil Brown
TEng, MIMGTechE, MIED
Editor: *MG Book Publishing*
Former Editor: *Whats New in Electronics* and *Electronics News*

Butterworths
London Boston Singapore Sydney Toronto Wellington

First published as *Dictionary of Electrical, Electronics and Computer Abbreviations,* 1985
Revised edition 1988

British Library Cataloguing in Publication Data

Brown, Phil, *1950–*
Electronics and computer acronyms.–
2nd ed.
1. Electronic engineering. Abbreviations
I. Title II. Brown, Phil, *1950–*
Dictionary of electrical, electronics and
computer abbreviations
621.381′0148

ISBN 0-408-02398-8

Library of Congress Cataloging-in-Publication Data

Brown, Phil (Phil Raymond)
Electronics and computer acronyms/Phil Brown. – Rev. ed.
p. cm.
Revision of: Dictionary of electrical, electronics, and computer abbreviations.
Bibliography: p.
ISBN-0-408-02398-8 :
1. Electric engineering–Acronyms. 2. Electronics–Acronyms. 3. Computers–Acronyms. I. Brown, Phil (Phil Raymond). Dictionary of electrical, electronics, and computer abbreviations. II. Title.
TK9.B76 1988
621.381′0148–dc19 87-37489

Photoset by Butterworths Litho Preparation Department
Printed in Great Britain by Robert Hartnoll (1985) Ltd., Bodmin, Cornwall

Preface to the new edition

The first edition of this volume, I am pleased to say, found wide acceptance not only in western countries but also in the Far East. However, technology has advanced at an incredible rate since its publication, and naturally this dictionary reflects that progress. This in itself poses a problem, as many companies are now continually coining acronyms. Some are never used used while others find wide acceptance and become almost industry standards. Constant vigilance is therefore maintained in order to sustain this volume's reputation, consequently there is a 'lead and lag' between inception and acceptance of any term, as there is between a manuscript and its final publication. Despite this fact, every effort has been made to keep the information contained herein current.

In this revised second edition, now called *Electronics and Computer Acronyms* – a title which aptly reflects the nature of its contents – you will find nearly 400 new entries, bringing the grand total to almost 2500 separate definitions.

This expansion of the electronics and computer emphasis is typified by entries such as MS DOS, GEM and WYSIWYG reflecting the computer related enhancements, while ASIC, BIST and EBL reflect those related to electronics.

P.R.B.
Shooters Hill, 1988

Preface to the first edition

Abbreviations can communicate information and provide understanding of complex ideas succinctly and swiftly. Their derivation, however, can be from numerous sources, brought about by the curtailment of principles during general usage or by the 'coining' of labels at the initial or conceptual stage.

Though the principles of abbreviations might appear simple to those who use them, their derivation can be haphazard because there are no specific rules or laws governing the production of acronyms or abbreviations – either they are born or they evolve. For example, the 'dual-in-line' concept has been logically curtailed to DIL, so that items are referred to as 'DIL sockets'; but when the housings or packages employed in the DIL system are used they are not referred to as 'DILP items' (dual-in-line packages) as one might expect but are termed 'DIP items'. Therefore, guesswork when employing or trying to understand abbreviations can lead to misunderstanding in communication.

The material for this book has been drawn from a number of subject areas, including: electrical, electronics, computers, telecommunications, fibre optics, microcomputers/microprocessors, audio, video, information technology, avionics, military, data processing, instrumentation, units, measurement, standards, services, organizations, associations and companies.

The mix has been selected to provide a comprehensive and broad view of 'electronics' and all that is associated with it, and as there is now generally an overlapping of disciplines (e.g. fibre-optic terms in data processing, computers within electronics) this dictionary can assist as an *aide-mémoire* when reading specification sheets and other technical publications, and when keeping up-to-date with institution journals and trade magazines.

Cross-references have been provided in the form of abbreviations within parentheses. For example, the entry for CCD, charge-coupled device, provides three cross-references: charge-transfer device (CTD), metal-oxide semiconductor (MOS), and capacitor (C,6). To broaden further the explanation, if required, referral to entries CTD, MOS and C part 6 (C is a multiple entry with ten parts) is advised. The system of prefixes and units used in the Système International d'Unités has been followed throughout this book. For further clarification, if required, the reader is referred to the entry SI,1.

For the engineer or technician this dictionary will serve as a memory jogger which involves them to use the cross-references to obtain a greater understanding of the subject. Whilst for sales and purchasing personnel, use as a reference companion to keep abreast of commonly-used terms is recommended as the most rewarding and time-saving employment of this book.

If further information is required the following publications are recommended:

Dictionary of Audio, Radio and Video by R. S. Roberts
Dictionary of Data Processing (2nd ed) by Jeff Maynard
Dictionary of Electronics (2nd ed) by S. W. Amos
Dictionary of Electrical Engineering (2nd ed) by K. G. Jackson, revised by R. Feinberg
Electronics Engineer's Reference Book (5th ed) by F. Mazda

P.R.B.
Shooters Hill

A

a Atto. A decimal prefix commonly used in association with a base unit in Le Système International d'Unités (SI) system of units, indicating a multiplication factor of 10^{-18} of that unit. For example, one attosecond, as, is equal to 0.000 000 000 000 000 001 of a second.

A (1) Ampere. The unit of electric current in the Système International d'Unités (SI) system of units named after the French scientist André-Marie Ampere (1775–1836). It is the electric current (I) which, when flowing in two infinitely-long parallel conductors, with their centres 1 metre (m) apart in a vacuum, causes each conductor to experience a force of 2×10^{-7} newtons per metre length of conductor.

(2) The hexadecimal (HEX) symbol for binary 1010, decimal 10. $A_{16} = 10_{10} = 12_8 = 1010_2$.

(3) Anode. (a) The electrode from which a current (I) flows, with the convention that the current direction is defined in terms of the movement of positive charges. (b) In an electron tube, the anode is the output electrode of the tube.

Å Ångstrom. A unit of length equal to 10^{-10} m. It is often used to express the wavelengths of very-high-frequency signals such as light.

AA One of the letter designations given to the most commonly-used sizes of 1.5-volt (V,1) batteries. Originally of North American nomenclature, this type of letter description is gaining increasing popularity over the International Electrotechnical Commission (IEC) letter/number system as it is able to simplify the selection of batteries. *See* battery-size cross-reference guide, Appendix.

AAA Another of the letter designations given to the most commonly-used sizes of 1.5-volt (V,1) batteries. *See* AA above.

AB (1) AB pack. A term used in the USA to describe a complete portable power source for electron tubes, comprising the A battery supplying the heater power and the B battery supplying the anode circuit power.

(2) Automated Bibliography. A history and description of a number of books, a list of documents pertaining to a given author or an annotated catalogue of documents held in a computer file.

ABA American Bankers' Association. An association that

promotes help or assistance for its members. An example is ABA coding; numbers that assist computerized cheque clearance.

ABC Automatic Brightness Control. A circuit used to adjust automatically the average luminance (brightness) level of a display to compensate for changes in ambient light level.

ABCA America, Britain, Canada and Australia Standard. Representatives of the four governments participate in the meetings of this body to agree on standards, especially those applying to military suppliers and industry.

ABEND ABnormal END. An irregular or freakish termination to a job or computer operation following either an error condition or an intervention by the operator.

ABIST Autonomous Built-In Self Test. A concept of the addition, through software, to very-large-scale integration (VLSI) devices, of the ability to carry with them the facility to self test, thus aiding automatic test equipment (ATE). ABIST differs from built-in self test, in that the entire program can be automated.

ABL Atlas BASIC Language. A computer language designed for use on an Atlas computer. *Also see* BASIC.

ABR Automatic Baud Recognition. A system in which a data communications channel (CHNL(c)) determines the data rate of the incoming data and adapts itself to that speed. *Also see* BAUD.

ABS (1) Acrynolite Butadine Styrene. A mouldable plastics compound that provides a rigid, tough and scratch-proof material.

(2) ABSolute value. An arithmetic function in beginners all-purpose symbolic instruction code (BASIC) and other computer languages that determines the magnitude or modulus of x, i.e. (x).

Examples:

ABS (11.9) = 11.9

ABS (−10) = 10

AC (1) Alternating Current. An electric current (I) that has its direction in a circuit (CCT) reversed with a frequency (*see* Hz) that is independent of circuit constants. Alternating currents are normally represented by a sine wave, as shown in *Figure A.1*.

(2) ACcumulator. *See* ACC.

ACARD Advisory Council for Applied Research and Development. A body in the UK that provides the government with advice concerning research and development (R & D) policies, as well as publishing reports.

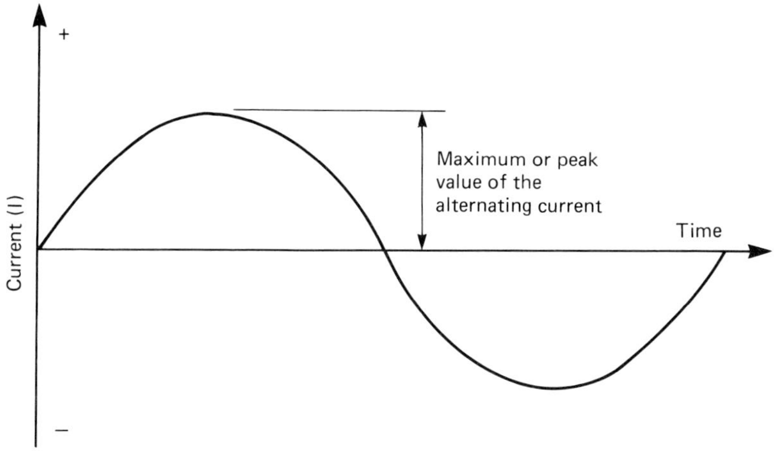

Figure A.1 One complete cycle of an alternating current (AC,1)

ACC ACCumulator. (a) A secondary cell having a reversible chemical action and charged by passing a current (I) through it. Direction and rate of chemical action are determined by the value of the external voltage (V,2). (b) A register of the arithmetic/logic unit (ALU) of a central processor unit (CPU) used as intermediate storage during the formation of algebraic sums, or for other intermediate logical and arithmetic operations.

ACE Automatic Computing Engine. A name given to one of the earliest computers.

ACIA Asynchronous Communications Interface Adaptor. A device that provides the data formatting and control to interface serial asynchronous data communications information to bus organized systems (i.e. between a microprocessor and a modulator/demodulator (MODEM)).

ACK ACKnowledge. A character code indicating a positive acknowledgement that a message has been received correctly, or that a receiver is ready to receive data.

ACL Advanced Cmos Logic. The latest step in the development of complementary metal-oxide semiconductor (CMOS) technology. Device speeds match those of bi-polar transistor-transistor logic (TTL) products in many cases, but with lower CMOS power dissination levels. The low power consumption makes ACL potentially more reliable than bipolar logic, making ACL devices particularly suitable for computers and peripherals, telecommunication devices, portable equipment and military applications.

ACM Association for Computing Machinery. A professional and technical society whose activities, publications and conferences are designed to aid computing advancement. The aid is with specific regard to machinery and system design, language and program development, plus other related activities.

ACOMPLIS A COMPuterized London Information Service. Held on a computer, this was an information service run by the Greater London Council's Library Department.

ACRE Automatic Call-Recording Equipment. An invention of British Telecom (BT,1) which dispenses with the need for operator written 'tickets' detailing calls they handle. ACRE records the details automatically on magnetic tape for subsequent processing by BT's computer-based telephone billing system.

ACTP Advanced Computer Technology Project. Funded by government in the UK, this is a project for research into and development of advanced computer techniques for industry.

ACU Automatic Calling Unit. A dialling device that permits a business machine to dial calls automatically over a communications network.

A/D Analogue-to-Digital. (a) The process of converting between an analogue and a digital signal. (b) A circuit used to convert information into digital form from an analogue input. Such a circuit is used in digital voltmeters and similar devices. *See* ADC, analogue-to-digital converter.

ADA Lady ADA Augusta Lovelace. A computer programming language named after Lady Ada (probably the first programmer) which is rigorously controlled by the US navy to avoid proliferation of dialects.

ADAPSO Association of DAta Processing Service Organisations. A data processing (DP,1) organization that also includes a software development and marketing group.

ADC or **A/DC** Analogue-to-Digital Converter. A device or circuit that changes analogue input voltages (V,2) to their equivalent digital (D,4) (binary or binary-coded decimal (BCD)) values for acceptance by the memory of a digital processor. An 8-bit ADC generates an 8-binary digit (BIT) word; a 12-bit ADC generates a 12-bit word. The encoding becomes more precise as more bits are used.

ADCs fall into three main groups, (a) integrating converters, (b) director flash converters and (c) feedback converters with internal self-checking digital-to-analogue converters (DAC) as shown in *Figure A.2*.

ADCCP Advanced Data Communications Control Procedures. A version of synchronous data link control (SDLC) standards

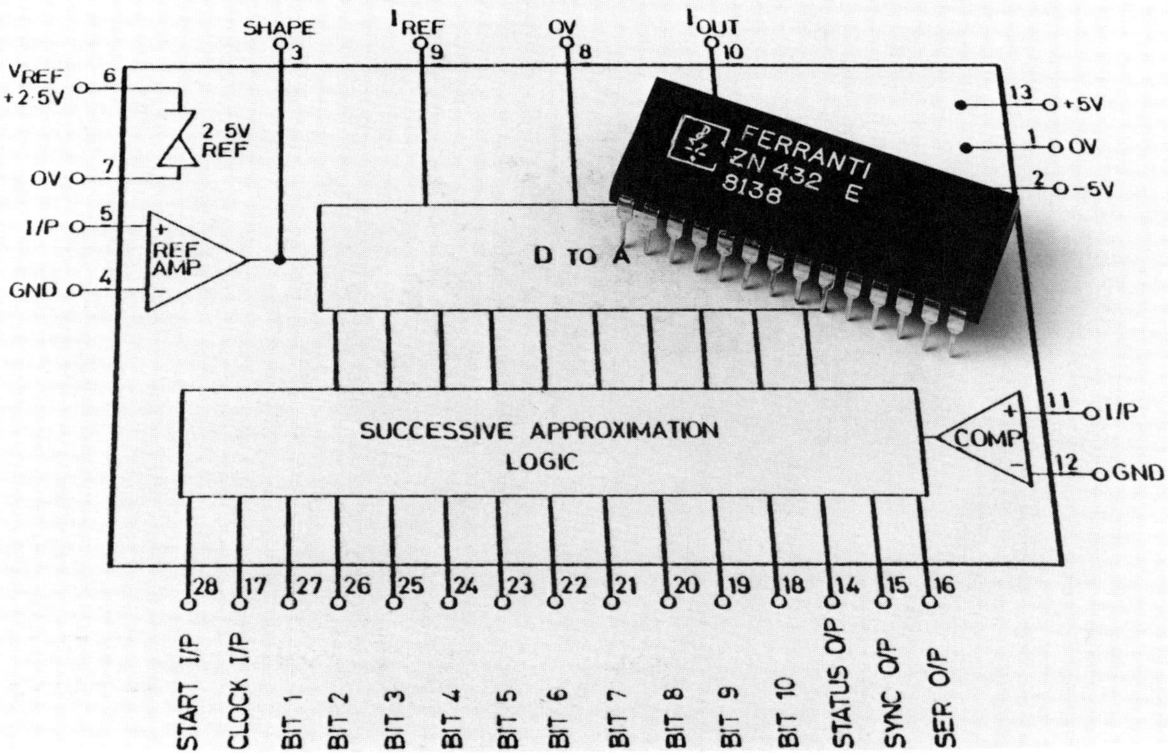

Figure A.2 Ten-bit analogue-to-digital converter (ADC) in a 28-pin package (*courtesy* Ferranti Electronics Ltd.)

produced by the American National Standards Institute (ANSI).

ADCON ADdress CONstant. A value or expression used to calculate the real or virtual storage (VS) address of a memory.

ADDCL ADD CLause. A learn as you go aid in the programming in logic (PROLOG) and other computer programming languages in the form of a built-in command. It adds new clauses to the memory of the computer (answers from previous questions or deductions) for future reference. Note delete clause (DELCL).

ADF Automatic Direction Finder. Typically an ADF consists of two crossed ferrite-core loops projecting 25 mm from the skin of an aircraft with an electronic goniometer generating a direct-reading display of bearing to the selected transmitter. For preference they are tuned to inland non-directional beacons (NDB), which exhibit a sharp reduction in signal strength when directly overhead.

ADLC Advanced Data Link Controller. Data that is transmitted and received in a synchronous serial form in a data communications system is converted into parallel form, analysed and stored by this equipment, so that data link management can be accomplished.

ADP (1) Automatic Data Processing. (a) Equipment such as electronic accounting machines and electronic data processing (EDP) units or systems. (b) Data processing (DP,1) performed by a system of electronic or electrical machines interconnected and interfacing so as to reduce to a minimum (MIN) the need for human assistance or intervention.

(2) Automatic Die Positioner. An item of production equipment that automatically selects and aligns good dies on a silicon wafer. The dies are collected by an automatic mechanical probe and bonded to the substrate of a chip (CHIP) package.

ADPCM Adaptive Differential Pulse Code Modulation. A coder-decoder (CODEC) used in cordless telephones (CT).

ADPE Automatic Data Processing Equipment. Machinery employed to perform automatic data processing (ADP,1(b)) functions.

ADPS Automatic Data Processing System. A system employed to perform automatic data processing (ADP,1(b)) functions.

ADRS Automatic Document Request Service. Provided by the British Library Automated Information Service (BLAISE), this service permits subscribers at an on-line terminal to ask for the loan of, or copies of, documents from the British Library Lending Division.

ADSR Attack, Decay, Sustain and Release. The constituent parts of the amplitude of the envelope of the sound generated by a musical instrument; also applies to a computer with sound-producing facilities. Attack is the immediate rise; decay is the settling down; sustain is the steady level; and release is the falling away to silence.

ADU Automatic Dialling Unit. A device capable of automatically generating telephone dialling digitizer signals.

Ae AErial. A system of conductors employed at a radio or television (TV(a)) installation to radiate or receive electromagnectic waves. Now known as an antenna.

AEC Atomic Energy Commission. The first atomic energy commissions were constituted in countries that either had or were developing atomic capability i.e. US, UK and USSR. They were swiftly joined by France, Canada and India. From the start their aim was the prospect of limitless cheap electricity, i.e. the use of the atom for peaceful means.

AECMA Association Européene des Constructeurs de Material Aerospatial. Concerned with aerospace matters, this association prepares draft standards that are subject to normal European Committee for Standardisation (CEN) procedures before becoming adopted as European Standards.

AES Auto-Electronic Selector. An all-electronic system which automatically connects a source (record player, tuner, cassette deck) through to the preamplifier when the source is operated.

AF Audio Frequency. Any frequency that can be detected by a normal human ear. This audile range extends from about 20 to 20 000 Hz. In a communication system intelligible speech can be obtained if a frequency range from about 300 to 3400 Hz is reproduced; any frequency in this range is called a voice frequency (VF).

AFC Automatic Frequency Control. A device that automatically maintains the frequency of an alternating voltage within specified limits. Error-operated, the device normally consists of two parts, a frequency discriminator that compares actual with desired frequencies, and a reactor controlled by the discriminator in such a way as to correct any frequency error. *See Figure A.3.*

AFID *See* AFIPS.

AFIPS American Federation of Information Processing Studies. An association of various USA data processing (DP,1) groups, institutes, societies and organizations. Formerly 'AFID'.

AFNOR Association Français de NORmalisation. A standards authority in France that is a member of the International Standards Organisation (ISO).

AFSK Audio Frequency-Shift Keying. A system employed in

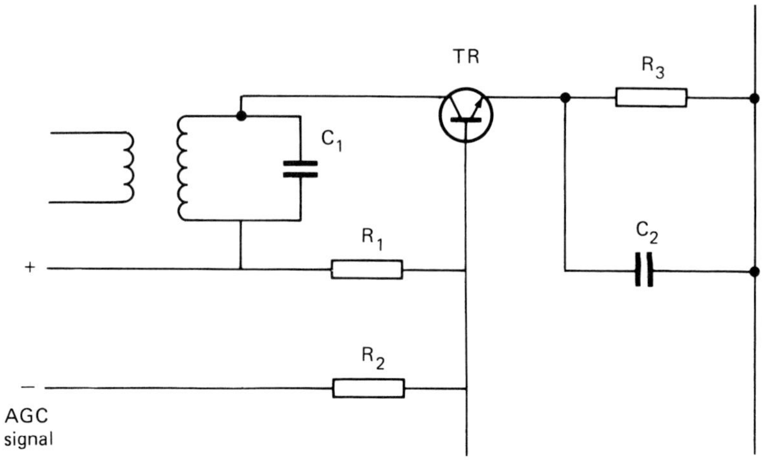

Figure A.3 Automatic frequency control (AFC) undertaken by a typical intermediate frequency (IF,1) stage

telecommunications that uses audio frequency (AF) tones to modulate a carrier wave (CW,1) in order to convey a digital (D,4) signal.

Ag Argentum. Silver. A white precious metal that is an excellent electrical conductor. Atomic number, 47; atomic weight 107.880.

AGC Automatic Gain Control. (a) A device that maintains the output (O/P) volume of a radio receiver at a constant level, despite fluctuations in the input (I/P) signal. An AGC is usually applied to amplifiers. (b) The process of maintaining the output volume of a radio receiver at a constant level.

AGR Advanced Gas-cooled Reactor. The particular type of reactor chosen by the British Central Electricity Generating Board (CEGB) as a second-generation follow-up to the Magnox reactor.

AI Artificial Intelligence. (a) A particular branch of computer science begun in the mid-1950s which endeavours to produce machines capable of exhibiting behaviour that if performed by humans would be considered intelligent.
(b) The attribute exhibited by some computers.

AIM Avalanche-Induced Migration. An irreversible process employed in the programming of a programmable read-only memory (PROM), based on aluminium migration through silicon to short circuit a diffused junction. *See Figure A.4.*

AIO Action Information Organization. An operations room on a warship that takes inputs from aircraft communications, radar

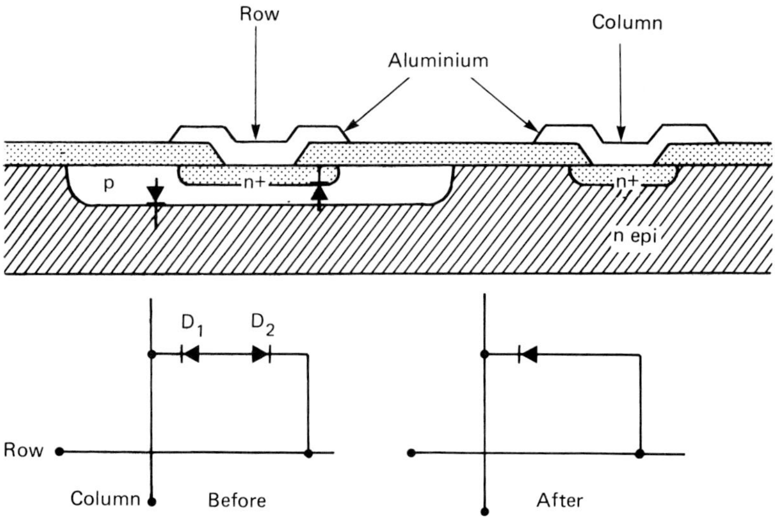

Figure A.4 The basic structure of an avalanche-induced migration (AIM) cell

(RADAR), sonar, computerized gun and missile controls and internal communications. Such information is displayed on consoles so that necessary human or computer intervention can be made, for instance to take evasive action when under attack.

AKWIC Author and Key Word In Context. An index that lists available computer programs or documents tabulated in alphabetical order of the most informative or significant word in the title of the information, along with the writer or programmer of the particular item of data.

ALC Automatic Level Control. A cassette tape recorder facility for the automatic adjustment of recording level.

ALD Automatic Location Device. A cassette tape recorder facility for finding the start of a specific track.

ALGOL (1) ALGOrithmic Language. A computer language by which numerical procedures may be precisely presented to a computer.
(2) ALGebraic Orientated Language. The international procedural computer programming language used especially for scientific applications.

ALOFT Airborne Light Optical Fibre Technology. A system used for the economic analysis of fibre optics technology in the American A-7 aircraft.

ALPHAMERIC A contraction of 'alphanumeric'.

ALS TTL Advanced Lower power Schottkey Transistor-Transistor Logic. A recent addition to the transistor-transistor logic (TTL) range of devices, the ALS TTL employs improved integrated circuit (IC) construction techniques, permitting smaller geometry on the integrated circuit slice, which provides higher speeds for no power increase. It is available with a propagation delay of 4.5 nanoseconds (ns) and a power consumption of 1.5 milliwatt.

ALU Arithmetic and Logical Unit. The register or portion of hardware of a computer or processor in which binary arithmetic and logical functions are performed, as shown in *Figure A.5.*

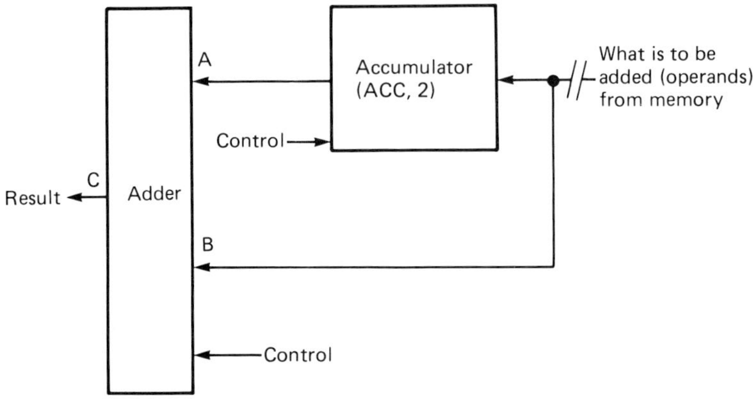

Figure A.5 One particular type of arithmetic and logical unit (ALU)

AM Amplitude Modulation. (a) A modulation in which the peak values, positive and negative, of an alternating current (AC,1) or wave are varied above and below their unmodulated values by an amount proportional to the peak value of the signal wave and at the frequency of the modulating signal, as shown in *Figure A.6.* (*b*) Also refers to a radio receiver or part of a radio receiver that detects amplitude-modulated signals only.

AM0 Air Mass ZERO. Also known as the 'solar constant', AM0 is generally used in relation to solar energy cells. It can be described as the radiant power per unit area perpendicular to the direction of the Sun outside the Earth's atmosphere, but at the mean Earth-Sun distance. The radiation intensity is referred to as the AM0 radiation.

AM1 Air Mass ONE. A term used in connection with solar cells. When the Sun is directly overhead, the length of its path through the atmosphere is at its shortest; thus the optical air mass is considered unity and the radiation is described as AM1.

AMACUS Automated Microfilm Aperture Card Update System. A system employing aperture cards, consisting of an 80-column

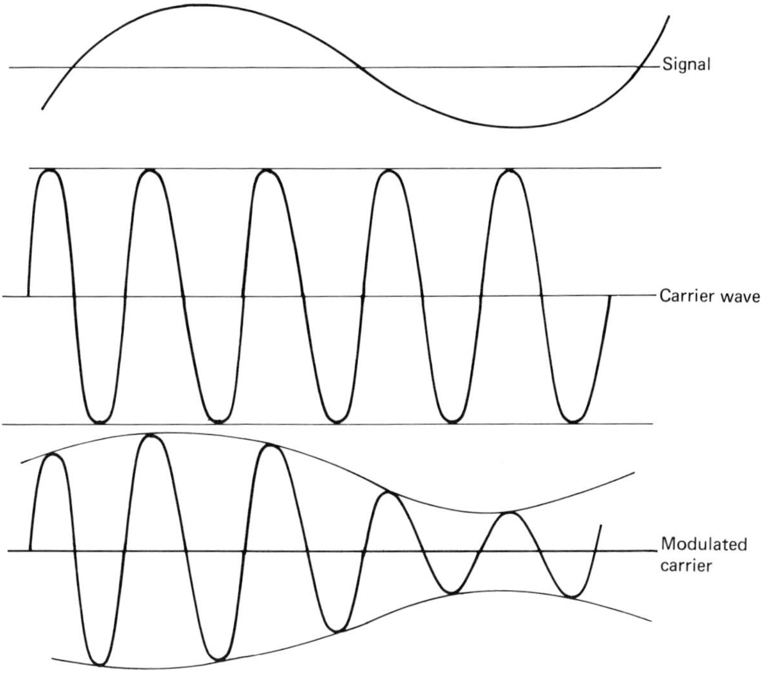

Figure A.6 Amplitude modulation (AM,(a))

punched card with a 35 × 45 mm frame of microfilm inserted, as its main form of storing information.

AMBIT Algebraic Manipulation BIT. A particular computer programming language used for algebraic symbol manipulation.

AM/FM Amplitude Modulated/Frequency Modulated. Usually refers to a radio receiver that detects both amplitude modulated (AM) and frequency modulated (FM) signals.

AMIE Advanced Manufacturing In Electronics. A programme aimed at raising the level of awareness of the changing technology in the electronics industry in the UK. It is designed to give a balanced view of the whole production process from concept to delivery. The programme is managed for the Department of Trade and Industry by the Institution of Electrical Engineers (IEE).

AMP (1) AMPlifier. A circuit (CCT) device or stand alone unit that is capable of increasing the magnitude of a given electrical input signal employing energy taken from an external source (i.e. it introduces gain). There are many types of amplifier and the particular nomenclature is dependent upon their applica-

tion, construction and whether they are operated by an alternating current (AC, 1) or a direct current (DC, 1). But basically there are two main groups: linear amplifiers (where the output (O, 2) signal is a linear function of the input signal) and non-linear amplifiers. *See Figure A.7*
(2) A contraction of the word Ampere. *See* A.

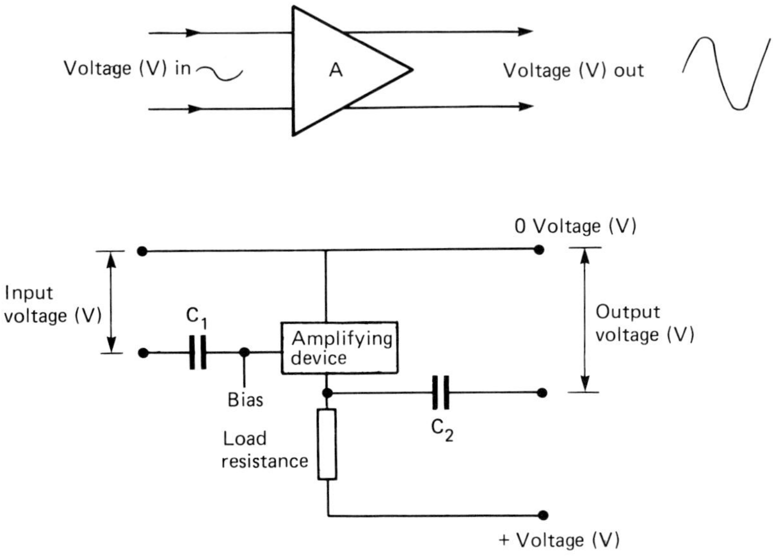

Figure A.7 Schematic diagram of an amplifier (AMP) and a typical amplifier circuit (CCT)

AMPS Advanced Mobile Phone Service. A cellular radio system developed by the Bell Telephone Company, which first emerged for trials in 1978 and is the forerunner of current cellular radio systems.

AMSAT The Radio AMateur SATellite Corporation. A corporation formed in the USA in 1969 to co-ordinate a range of American and international amateur space projects. *See* OSCAR.

AND A logic function, operation, circuit or gate. (a) A logical function or operation defined by a specific rule that if two statements, A and B, are true (1), then C is also true (1); if not, then C is false (0). Truth is usually expressed by the value 1, while 0 is used to indicate a false state, as shown in *Figure A.8*. (b) A physical circuit that performs the AND function. (c) A gate that performs the AND function.

MIL-STD-806B	BS 3939 ANSI Y.32.14-1973 IEC 117	Old BS 3939	Sometimes used symbol	Truth table

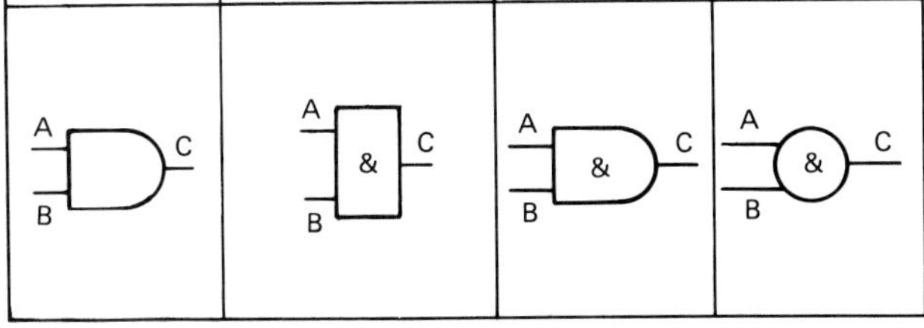

A	B	C
0	0	0
0	1	0
1	0	0
1	1	1

Figure A.8 The AND gate. In practice the MIL-STD-806B symbol is the one most likely to be met, although the other logic symbols may be encountered.

ANL Automatic Noise Limiter. A circuit employed in radio transmitter/receivers, now extensively used in citizen band (CB,1) transceivers to reduce noise or interference such as that caused by motor-car ignition systems.

ANSI American National Standards Institute. An organization that establishes standards and organizes committees of computer users and manufacturers. It develops and publishes industrial standards in the fields of computing and information handling that are accepted world wide. Formerly USASI.

A-O Acousto-Optic. A term that refers to the interaction of optical and acoustic waves.

APD Avalanche PhotoDiode. A photo-detecting diode (D,5) sensitive to light energy. It increases its electrical conductivity by increasing the number of electrons in its conduction band through absorbing photons of energy, the interaction of electrons, and an applied bias voltage (V,2).

APG Automatic Program Generation. A self-contained package for the generation of test code statements.

APL Algorithmic/Procedural Language. A programming language with an extensive set of operations and data structures that are used to implement what is considered by many to be the most flexible, powerful and concise language in existence. It was invented by Kenneth Iverson and is also known as 'A Programming Language'.

APT Automatically Programmed Tools. A system for the computer-assisted programming of numerically controlled (N/C) machine tools, draughting machines and similar equipment.

AQL Acceptable Quality Level. Batches of components or parts are inspected and laboratory tested by many manufacturers to a specific percentage value. If more than that percentage fail to meet the AQL, the whole batch is rejected.

ARPA Advanced Research Projects Agency. An agency of the US Department of Defense, which supports its own resource-sharing computer network (ARPANET).

ARPANET Advanced Research Projects Agency NETwork. A large packet-switching network that employs exchanges linked by high-speed data-transmission facilities, developed by the US Department of Defense.

ARQ Automatic ReQuest for repetition. An automatic system that provides error correction by making use of a code and a construction such that any false signal initiates a repetition of the transmission of the character incorrectly received. Also known as 'automatic error correction'.

ARR Automatic Retransmission Request. A technique that permits a system to initiate an automatic request for retransmission when a data error is detected in a received message.

ARTEMIS Automatic Retrieval of Test through European Multipurpose Information Services. Currently under evaluation by the European Commission, ARTEMIS is planned to be a system that can deliver documents, supplying articles of text on demand over a facsimile (FAX) or digitized text-transmission system.

ARU (1) Audio Response Unit. A device that can connect a computer system to a telephone and thereby provide a voice response to any enquiries received.

(2) Acoustical Resistance Unit. A device of predetermined acoustical resistance juxtaposed with the port opening of a reflex loudspeaker. It controls the Q-factor (*see* Q-factor) or frequency response.

ASA American Standards Association. A US association responsible for the establishment of standards; formerly known as the United States of America Standards Association.

ASC Automatic Sequence Control. A management feature found in computer programs.

ASCC Automatic Sequence Controlled Calculator. An early computer constructed by Howard Aiken at Harvard University in 1937.

ASCII American Standard Code for Information Interchange. Usually pronounced *ask-ee*. This is an 8-binary digit (BIT) information code for data transfer used with most computer and data terminals. It has been adopted by the American Standards Association (ASA) to achieve compatibility between data devices and other pieces of equipment.

ASIC Application Specific Integrated Circuits. Gate arrays and standard cell arrays layed out for use in specific applications primarily for specific customers. *See* integrated circuits (IC).

ASP Attached Support Processor. A term used to describe multiple computers, usually two, connected via channel-to-channel adaptors, employed to increase efficiency in processing many short duration tasks.

ASR Automatic Send-Receive set. A combination teletypewriter (TTY), transmitter and receiver having the capability to transmit from either a keyboard (KB,1) or paper tape (PT).

ASS Automatic Selector Switch. A microswitch operated system which automatically connects a source (record player, tuner, cassette deck) through to the preamplifier when the source is operated.

ASW Anti-Submarine Warfare. A typical acoustic processing and display advanced airborne anti-submarine defence system comprising receivers, digital (D,4) data processors, operator controls and the displays (D,6) on which information about a submarine is presented.

ATE Automatic Test Equipment. Equipment or systems that can test devices automatically. For complex devices, especially semiconductor devices, the systems are usually programmable. They may be purpose-built or constructed from units employing the general purpose interface bus (GPIB) standard. Operators are needed only to connect the unit under test (UUT) to the system.

ATLAS Automatic Tabulating, Listing And Sorting system. A package of computer software that is used for tabulating, listing and sorting.

ATM Automated Teller Machine. Automatic machines or terminals that are activated by a banker's card and the entry of a personal identification number (PIN), *See Figure A.9.* They provide money (bank notes) on request, via a visual display unit (VDU) and a keyboard (KB,1) through a sequence of questions and replies, 24 hours a day.

ATR Anti-Transmit-Receive. A switch that automatically decouples a radar (RADAR) transmitter from its antenna during a receiving period; extra protection is thus provided from return echoes received by the antenna. *See* (TR) Transmit-Receive.

ATS Application Technology Satellite. Used in particular for medical communications, this is the designation given to a series of American communications satellites that provide both audio and video channels (CHNL(g)).

ATTD Avalanche Transit-Time Device. The generic name given to a family of negative resistance microwave (μW) oscillator diodes (D,5). The family includes barrier injected transit-time (BARITT) diodes, impact ionization avalanche transit-time

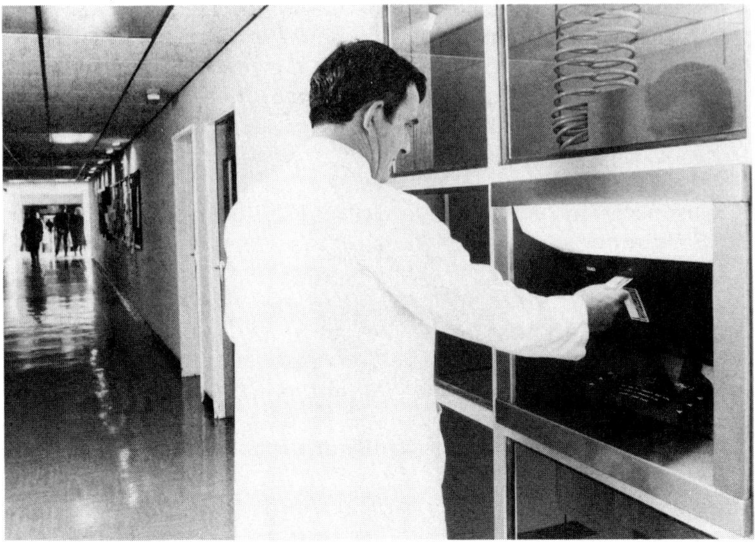

Figure A.9 An NCR 5080 ATM (automated teller machine) installed in a factory for staff use. (*courtesy* NCR Ltd.)

(IMPATT) diodes, and trapped plasma avalanche triggered transit (TRAPATT) diodes (D,5).

ATU Antenna Tuning Unit. A specific circuit that adjusts and varies the combined effects of capacitance (C, 3), resistance (R,1) and inductance of an antenna to match that of transmitters and receivers.

AUTO AUTOmation generation. A command in the common algorithmic language (COMAL) computer programming language which provides for the automatic generation of line numbers in a program.

AUTOABSTRACT AUTOmatic ABSTRACT. Significant words from the title and sometimes the body of a document that describe its contents are selected and arranged in a meaningful pattern. The arrangement and selection by computer is carried out according to a specific programmed standard. Although not looking like an abstract prepared by humans, the autoabstract normally provides a good indication of what the document is about.

AUTOPOLL AUTOmatic POLLing. Without interrupting the central processing unit (CPU), this hardware feature is able to process a list specifying the order in which a number of computer terminals are polled. It also handles negative responses.

AUTOVERIFIER AUTOmatic VERIFIER. Equipment employed to read information punched on a card and verify that information.

AUX AUXiliary. An input made available on an amplifier (AMP,1) (or receiver) for the connection of an additional sound source such as a cassette deck or compact disc (CD,2) player.

AV A chrome dioxide video format developed by Sony.

AVC Automatic Volume Control. A term that is still in use but should be deprecated in favour of 'automatic gain control'. (*See* AGC(a))

AVPO Axial Vapour-Phase Oxidation process. A process used for making graded-index (GI) optical fibres (OF) in which the glass preform is grown radially rather than longitudinally as in other processes.

AVR Automatic Voltage Regulator. A circuit or device that maintains the terminal voltage (V,2) of a generator or other voltage source within prescribed limits, despite variations in input, voltage or load.

AWAC Airborne Warning And Control. A powerful radar (RADAR) system mounted on a Boeing aircraft to maintain a continuous watch on possible hostile missiles and aircraft from above. It can detect those ground-hugging aircraft and missiles that remain unseen by ground-based radar. AWAC was selected by the UK Government in preference to the GEC Avionics Nimrod in December 1986.

AWG American Wire Gauge. A system of gauging or sizing wires that provides a particular number as a designation for a specific diameter of wire and is frequently employed in the USA. It is also known as Brown & Sharpe gauge (B & S).

AWM Appliance Wiring Materials. Used as a prefix thus; AWM UL, this designation refers to particular sheathings for cables and wires are recognized by underwriters laboratory (UL).

AZERTY A French language computer terminal or typewriter keyboard (KB,1) which has the keys arranged such that the top row below the numerals begins with the letters, A, Z, E, R, T and Y. *See Figure A.10.* It normally contains some additional letters peculiar to the French language, as well as instructional keys notated in French. *See* QWERTY and QWERTZ.

Figure A.10 An AZERTY keyboard (*courtesy* Alphameric Keyboards Ltd.)

b Base of transistor. An area or region in a bipolar transistor between the emitter (e) region and the collector (c,2) region, into which carriers are injected from the emitter region.

B (1) Battery. Nowadays usually the B part of the AB pack that supplies power to the anode circuit of an electron tube. In the early days of radio the B battery was a multi-cell Leclanché A battery which was a lead-acid accumulator (ACC(a)). *See* AB,1.
(2) The hexadecimal (HEX) symbol for binary 1011, decimal 11. $B_{16} = 11_{10} = 13_8 = 1011_2$.

BAEC British Amateur Electronics Club. A national amateur electronics club in the UK that helps all who are interested in electronics. Started in 1966, it has many members from beginners to experts in all parts of the British Isles and overseas.

BAF Bonded Acetate Fibre. A material used for acoustical damping in some loudspeaker cabinets.

BAL Basic Assembler Language. A non-complex computer program that translates an assembly language into object code, having no macros or conditional assembly instructions.

BALUN BALanced to UNbalanced transformer. Typically this type of transformer (XFMR) is employed to couple a balanced antenna to an unbalanced transmission line.

BAM Basic Access Method. Each input/output (I/O) statement in this method of computer access initiates the performance of a machine input/output operation.

BARITT BARrier Injected Transit-Time. A microwave (μW) oscillator diode (D,5) which exhibits a lower noise figure than an impact ionization avalanche transit-time (IMPATT) diode. However, it has the disadvantage of a relatively narrow bandwidth and a lower power output (O/P) of only a few milliwatts.

B & S Brown & Sharpe gauge. A system of gauging or sizing of wires that provides a particular number as a designation for a specific diameter of wire and is frequently employed in the USA. Also known as 'American wire gauge' (AWG).

BASIC Beginners All-purpose Symbolic Instruction Code. A popular computer language characterized by its ease of use and the ease with which it can be learned. It is the standard language used with microcomputers and is well-suited to computer time-sharing.

BASIS British Airways Staff Information System. A videotex system employed by British Airways for conveying information.

BAT BATtery. A source of direct current (DC,1) or voltage (V,2), producing electrical energy by chemical means, it consists of one or more electrolytic cells connected together and used as a single unit.

BATS Basic Additional Teleprocessing Support. A system used by some computer systems in the late 1960s to provide telecommunication processing support.

BAUD BAUDot. A unit of measure of data signalling rate or data flow. It is the instantaneous bit speed with which a device or channel (CHNL(b)) transmits a character in serial transmission. Although strictly speaking it stands for 'signal speed', for practical purposes it is generally considered as interchangeable with bits per second (BPS). Emil Baudot produced the first widely-used teleprinter in the nineteenth century, as well as the Baudot code for use on telegraphic circuits.

BAVIP British Association of Viewdata Information Providers. A body comprised of individuals or organizations that provides information or data for viewdata systems.

BBC British Broadcasting Corporation. The independent non-commercial radio and television (TV(a)) authority in the UK, that is funded by the legally enforced licensing of television receivers.

BBD Bucket-Brigade Device. A device consisting of capacitors (C,6) linked by switches. The switches in practice are bipolar or metal-oxide silicon field-effect transistors (MOSFETs) as shown in *Figure B.1*. These circuits are now usually made as

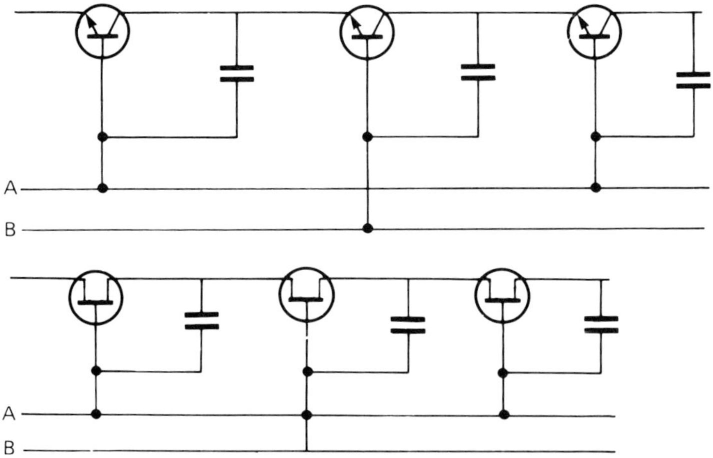

Figure B.1 Bucket-brigade devices (BBD): top employing bipolar transistors; bottom employing metal-oxide silicon field-effect transistors (MOSFET)

integrated circuits (ICs). Clocked or timed pulses are applied to close the switches, a two-phase system (A and B) being used. As each switch is closed charge is transferred from one capacitor to another.

B-BOX (a) A register that contains a quantity to be used under direction of the control section of computer hardware. (b) A device that permits automatic modification of an instruction address without permanently altering the instruction in the memory. (c) A register to an arbitrary integer, usually one which is added (or subtracted) upon the execution of each machine instruction.

BCC Block Check Characters. Check binary digits (BITs) used in line control procedures, which make up the later part of a block of transmitted information. They are generated by a checking process applied to the information part of the transmission.

BCCD Bulk-channel Charge-Coupled Device. A charge-coupled device (CCD) that has formed, by epitaxy or ion implantation, a thin layer on the surface of the substrate of an opposite conductivity type. It thus confines the flow of charges to a channel (CHNL(h)) lying beneath the surface of the semiconductor. Also known as a 'buried-channel charge-coupled device'. *Figure B.2.*

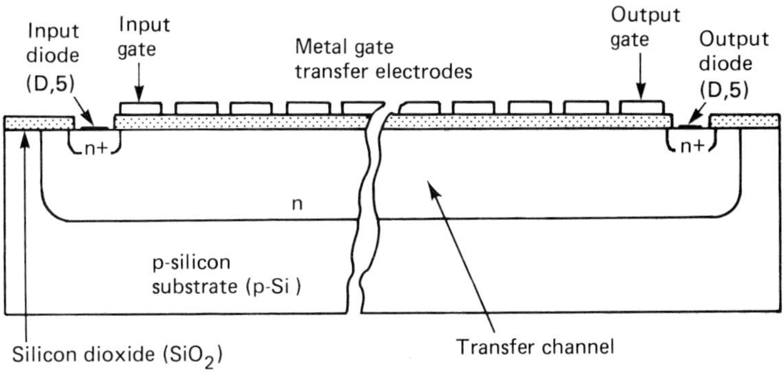

Figure B.2 A cross-sectional view of a bulk-(buried) channel charge-coupled device (BCCD)

BCD Binary Coded Decimal. A numerical representation in which digits (DGT,1) are represented by binary numerals (1 or 0), and the individual figures of a decimal number are separately coded into binary form. The most common code uses a presentation where 8-4-2-1 binary code groups are used, as shown in *Figure B.3.* In this code twenty three is represented as 0010 0011; in pure binary notation it would be 10111.

Binary	Decimal	Binary	Decimal
0000	0	0101	5
0001	1	0110	6
0010	2	0111	7
0011	3	1000	8
0100	4	1001	9

Figure B.3 Binary-coded decimal (BCD) numerical representation

BCO Binary Coded Octal. This is a system where binary numbers are employed to represent the octal digits (DGT,1) of an octal number.

BCPL Basic Combined Programming Language. Developed to write a computer program that translates from a high level into object code for the combined programming language (CPL), this system's implementation language is available on a wide range of machines.

BCS (1) Superconductivity theory. The BCS theory in which pairs of electrons (Cooper pairs) combine in the presence of other electrons. Interactions between electrons result in the pairing and vibrations of the crystal lattice providing a material with a highly-ordered state. The material then exhibits little resistance (R,1) to electron movement with no dissipation of energy.
(2) British Computer Society. Founded in 1957 and incorporated by Royal Charter in 1984, the BCS has over 27 000 members. It is internationally recognized as the professional organization devoted to computing in the UK and it conducts its own examinations which are professionally accepted throughout the world.

BD BackwarD. A command in the Logo computer programming language that instructs a Turtle graphics robot to move backwards a given distance.

BDAM Basic Data Access Method. Providing a non-queueing facility, BDAM is a method employed by some computers of gaining access to data held in a backing store. Although not a primary store the backing store has in fact a larger capacity but a slower access time.

BDOS Basic Disk Operating System. An area in the memory map (MAP,2) of a microcomputer using a control program for microcomputers (CP/M) disk operating system (DOS), having routines used by the system to access its system disk drives. It thus provides for file management.

BDPS Beam Deflection Phase Shifter. An enhancement of electron beam automatic test equipment (ATE), in which the electron beam passes through an aperture and is deflected around a rectangular path. It can be used to monitor internal waveforms of devices for fault diagnosis.

BDV BreakDown Voltage. The voltage (V,2) under specified conditions at which an equipment breakdown occurs. Compare BV which applies only to semiconductor devices.

BEEF Business and Engineering Enriched Formula translator. Employed in business and engineering applications, this is an enhanced version of formula translator (FORTRAN) computer language.

BEL BELl character. One of the range of characters that form a specific code for any particular system (character set) that will initiate the sounding of a bell on a terminal device.

BER Bit Error Rate. This is usually expressed as the number of erroneous binary digits (BITs) per million introduced in the passage of single or parallel streams of bits in a transmission path.

BEX Broadband EXchange. In the USA a public switched communication system of the Western Union Company. It features various bandwidth full duplex (FDX) connections.

BFL Back Focal Length. The distance measured from the vertex of the back surface of a lens to its rear focal point.

BFO Beat Frequency Oscillator. A circuit generating an alternating current (AC, 1) with an adjustable frequency. Its output (O/P) can be mixed with that of a final intermediate-frequency amplifier (AMP). It produces an audible frequency beat when the receiver is tuned to an unmodulated signal.

BH (1) Busy Hour. An uninterrupted period of one hour during which the average intensity of traffic of a telephone exchange or group of telephone circuits is at a maximum (MAX).
(2) Buried Heterostructure. A particular type of structure of stripe geometry semiconductor laser (*see* LASER).

BHJT Bipolar HeteroJunction Transistor. The heterojunction of a BHJT refers to the junction between two dissimilar semiconductor materials, one of which has a wider energy gap than the other. BHJTs have applications in low-speed power products, opto-electronic integrated circuits (IC), extremely high-speed discrete devices and logic circuits, and transistors that operate at extremes of temperature.

BIC Bipolar Integrated Circuit. A particular type of monolithic integrated circuit (IC,1), that has its basic structure of bipolar transistors.

BICMOS Bipolar Complementary Metal-Oxide Semiconductor. A technology that combines on a single chip (CHIP) both bipolar and complementary metal-oxide semiconductor

(CMOS) devices. These circuits are very compact and the technology eliminates many of the peripheral devices such as filters.

BIDAP BIbliographic DAta Processing program. A package of computer software designed to process information relating to books, articles or reports. Normally the data covered is of authorship, title and publication details.

BIFET BIpolar compatible Field-Effect Transistor. A junction-gate field-effect transistor (JUGFET) combined with bipolar transistors for some types of integrated circuits (IC) such as operational amplifiers (OP AMP).

billi BILLIon. A prefix to a measure value to indicate a billion units, e.g. billipounds. In the UK a billion has traditionally been a million millions (10^{12}), whereas in the USA and elsewhere it stands for a thousand millions (10^{9}).

BIM Beginning of Information Marker. A reflective spot about 3 metres (m,2) from the physical end of a magnetic tape (MAGTAPE). It is sensed by a photoelectric device and indicates the point on the tape at which recording may begin.

BIMOS BIpolar and MOS techniques. Techniques that combine on a single CHIP both bipolar and metal oxide semiconductor (MOS) devices. These high-density devices offer near-ideal operational-amplifier parameters and enable digital (D,4) and linear functions to exist on the same integrated circuit (IC).

BIN Basic Identification Number. The BIN code is a three-colour-band system on the contact body of avionic rectangular connectors to identify each contact size.

BINAC BInary Automatic Computer. A computer built by the Eckett-Mauchly Corporation in the USA in the late 1960s.

BIOD Bell Integrated Optical Device. An integrated optical circuit (IOC) having both active and passive elements, which can be used as logic and control elements in various other devices (e.g. logic gates) depending on how they are interconnected during fabrication.

BIOS Basic Input Output System. A group of routines comprising an area in the memory map of a microcomputer using a control program for microcomputers (CP/M) disk operating system (DOS), that is implemented when a user first gets the system up and running. The area is then called up by the operating system when it needs any input/output (I/O) routines.

BISAM Basic Indexed Sequential Access Method. A method employed by some computers to gain access to information held in a backing store, which is not a primary memory store. It is in fact a store that has a larger capacity but a slower access time.

BIST Built-In Self Test. Design features of very-large-scale integration (VLSI) semiconductor devices, that include test features as an aid to automatic test equipment (ATE).

BISYNC BInary SYNChronous communications. In a data communications system, BISYNC is the use of a set of control characters and sequences for the synchronized transmissions of binary coded data between stations. It is a low-level half-duplex (HD) communications protocol that allows any pattern of binary digits (BITs), including those normally used for control, to be transmitted in a block. Also called BSC.

BIT BInary digiT. The smallest amount of information, consisting of two alternatives, either 0 or 1. It is one character of a digital (D,4) word, and the position of the bit in the word usually determines its significance. For example, *Figure B.4* shows the column values compared with decimal numbers. The numbers can be obtained by progressing right to left across the columns and adding values when they appear; for example, 9 is 8+0+0+1, left to right.

Equivalent number (decimal)	Column values (decimal)				Binary digits (BITs)
	8	4	2	1	
0	0	0	0	0	
1	0	0	0	1	
2	0	0	1	0	
3	0	0	1	1	
4	0	1	0	0	
5	0	1	0	1	
6	0	1	1	0	
7	0	1	1	1	
8	1	0	0	0	
9	1	0	0	1	

Figure B.4 Binary digit (BIT) numerical representation

BITBLT BIT BLock Transfer. A feature that permits processors to move rectangular blocks of bit-mapped data.

BJT Base Junction Transistor. The centre semiconductor region of a double junction transistor, in which the output current is controlled by the input current.

BK BacKward. A command in the programming in logic (PROLOG) computer programming language that instructs a Turtle graphics robot to move backwards a given distance.

BLAISE British Library Automated Information ServicE. A computer service that contains the authorship, title and publication details of all books published in the UK since 1968.

Special software enables the information to be retrieved and edited to produce local catalogues.

BLAST Boolean Logic And State Transfer. Modules of design software that permit a designer to define pin neuronics and functions, and to write Boolean equations or state transfer equations which are compiled into a standard integrated fuse logic (IFL) device programming table.

BLO BLOcking signal. A signal transmitted to the distant end of a communication circuit to indicate that it has been reserved for use or has been rendered unavailable for use.

BNC Bayonet Nut Coupling. A type of connector used for coaxial cable (COAX) radio frequency (RF) interconnection. It employs a two-prong bayonet with a spring coupling unit. BNC connectors are extensively employed within test equipment as they provide a rapid connection/disconnection.

BNF Backus Normal Form. A formal computer language structure for the syntax phasing of other computer languages. Also known as 'Backus-Naur form'.

BNPF Beginning-Negative-Positive-Finish. An older type of data-encoding output format for programmable read-only memory (PROM) programmers. B represents beginning, N is negative (1), P is positive (O) and F is finish. For example, 101010 is represented by BNPNPNPF.

BOD Break Over Diode. A two terminal thyristor that is switched to the low impedance state by exceeding the forward breakover voltage (V,2). It is an undirectional device whose forward characteristic is similar to a reverse-biased avalanche diode (D,5).

BOP Bit Oriented Procotol. Usually refers to messages (i.e. BOP messages) used in telecommunications, that are transmitted in frames, all messages adhering to one standard frame format.

BOS Basic Operating System. Designed to run on small IBM computers, this software, held resident on a disk, is associated with the execution of programs and the co-ordination of a computer system.

BOSFET Bidirectional Output Switch Field-Effect Transistor. Produced by International Rectifier Co. in the UK, a BOSFET is a high-effeciency metal-oxide silicon field-effect transistor relay. It is designed to switch direct current (DC,1) voltages (V,2) from thermocouple levels to 100 volts (V,1) peak.

BOT Beginning Of Tape. The point to which a cartridge tape is rewound, as opposed to a beginning of information marker (BIM).

BPAM Basic Partitioned Access Method. A method employed

by some computers to gain access to information held in a backing store, which is not a primary memory store, but a store that has a larger capacity but a lower access time.

BPI Bits Per Inch. An expression of the density of data on a tape or disk memory-storage medium; the number of binary digits (BITs) over a specific length.

BPS (1) Bits Per Second. A unit of measure of data flow. It is the instantaneous binary digit (BIT) speed with which a device or channel transmits a character in serial transmission. For practical purposes the term is generally considered to be interchangeable with BAUD.
(2) Basic Programming Support. BPS card or tape systems are simple and small operating systems (OS) available for mainline computers.

BR Butyl Rubber. A material used for cable insulation. BR is a class of long chain polymer capable of being crosslinked to produce elastic compounds, i.e. an elastomer.

BRA British Robotic Association. An organization of all who are interested in industrial robots, including suppliers, potential users, government agencies and trade unions. It was founded in 1977 and is UK based.

BREMA British Radio Equipment Manufacturers Association. An organization that includes radio makers within the UK.

BROWSER BRowsing On-line With SElective Retrieval. A computer system data base able to offer ordinary English language searching of the documents that it holds.

BRS Bibliographic Retrieval Services. An American service offering access to over 20 of the most popular computer data files with information relating to books, articles or reports. Normally the information covered is of authorship, title and publication details.

BS (1) Back Space. One of the range of characters that form a specific code for any particular system (character set) that will cause the automatic backspacing of an input (I/P) or output (O/P) device.
(2) British Standard. A prefix normally used in conjunction with a number and a date to denote a particular standard laid down and published by the British Standard Institution (BSI). For example, BS1991: Part 6: 1975 refers to letters, signs and abbreviations for electrical science and engineering.

BS 9000 The British Standards 9000 System. A system of quality requirements and fundamental performance characteristics prepared for families of components such as capacitors (C,6), connectors, relays, resistors (R,2) and switches. Various levels are established by agreement between manufacturers, users, and National Supervisory Inspectorate (NSI) officers at the British Standards Institution (BSI). Approval tests can be authorized on behalf of the BSI by NSI officers at a local level.

BSAM Basic Sequential Access Method. A computer system with a large continuously updated file of information, abstracts or references concerning a particular subject (i.e. a data base).

BSC Binary Synchronous Communications. *See* BISYNC.

BSI British Standards Institution. An authority that establishes and publishes standards for measurement, nomenclature and product performance in the UK. BSI is also the UK member body of the International Standards Organisation (ISO).

BSN Back-end Storage Network. Normally having a high bandwidth, this network is able to connect a number of computers to high capacity memory storage devices.

BSS British Satellite Service. A satellite service aimed at community or large home receivers. These transmissions are at lower power than direct broadcast satellites (DBS), but higher than fixed satellite services (FSS).

BT (1) British Telecom. An autonomous entity responsible for telecommunications in the UK and dealing with services such as telephone, telex, telegraph, Datel and Prestel. In the last quarter of 1984 BT was privatized.

(2) Busy Tone. A signal conveyed to a telephone calling party to indicate that the called telephone party or intermediate apparatus is reserved for use or is rendered unavailable for traffic.

BTAM Basic Telecommunications Access Method. BTAM is used to control the transfer of data, in a computer system, between the main storage and local or remote terminals.

BTP Batch Transfer Program. Able to accommodate more than one terminal, a BTP controls data transfers from local or remote computer terminals, and allows for modification or deletion of records.

BTX BildschirmTeXt. The West German viewdata system.

BV Breakdown Voltage. The reverse bias voltage (V,2) at which a p-n (p-n) junction begins to conduct a large current (I). The junction will not be damaged unless excessive currents are allowed to flow. Compare with BDV which applies generally.

BW BandWidth. (a) A range of frequencies that can be occupied by a transmitted modulated signal. (b) The range of frequencies within which the performance of a device or system falls within given limits, with respect to a particular specified characteristic.

BYTE A colloquial expression to describe an 8-binary digit (BIT) word, in contrast to a 'nibble', which is generally considered to be a 4-binary digit word.

c (1) Centi-. A decimal prefix commonly used in the Système International d'Unités (SI) system of units, in association with a base unit of that system. Centi- denotes a multiplication factor of 0.01 or 10^{-2}.
(2) Collector of transistor. (a) An area or region in a bipolar transistor into which charge carriers are swept from the base region. The carriers are minority carriers in the base (b) region but majority carries in the collector region. (b) Also used to imply a collector of electrons e.g. an anode (A,3) in certain types of electron tube.
(3) Represents the speed of light in a vacuum. In the Système International d'Unités (SI) system of units, $c \simeq 3\times 10^8 \mathrm{ms}^{-1}$, (m = metres, s = seconds).

C (1) A high-level and structured computer programming language, which has been designed to optimize run time, size and efficiency.
(2) The hexadecimal (HEX) symbol for binary 1100, decimal 12.
$C_{16} = 12_{10} = 14_8 = 1100_2$
(3) One of the letter designations given to the most commonly used sizes of 1.5-volt (V,1) batteries. Originally USA nomenclature, this type of letter description is gaining increasing popularity over the International Electrotechnical Commission (IEC) letter/number system as it is able to simplify the selection of batteries. *See* battery-size cross-reference guide, Appendix.
(4) Candle. Now obsolete, the international candle was a unit of measurement of light intensity and has been superseded by the candela. 0.982 international candles = 1 candela.
(5) Capacitance. In a system of plates (conductors) and dielectric (insulators), capacitance is the ratio quantity of electric charge to the potential developed. It is measured in farads (F).
(6) Capacitor. A component consisting of two electrodes or plates separated by a dielectric (an insulator). It has appreciable capacitance ((5) above). The symbol C is used to represent a capacitor on parts lists and to annotate schematic diagrams.
(7) Carbon. A conductive non-metal which is also known as 'graphite' and has been used extensively in the electronics and electrical industries.
(8) Celsius. The suffix °C is sometimes taken to be degrees centigrade but now represents degrees Celsius. The two temperature scales are substantially the same, but with subtle differences. The centigrade scale is defined by 0° at the freezing

point of water and 100° at its boiling point. The Celsius scale is defined to be the same as the Kelvin (K,2) thermodynamic scale with the zero shifted to the ice point, which is 273.15 K.

(9) Coulomb. The unit of electrical charge in the Système International d'Unités (SI) system of units, it is equal to the quantity of electricity transferred by a current of 1 ampere (A) flowing for 1 second (s).

(10) Cathode. The electrode in an electron tube which is connected to the nagative supply and is the source for the principle electron stream.

CAAS Computer-Assisted Acquisition System. A computer system used to assist libraries in their acquisition of material.

CAD (1) Computer-Aided Design. Refers to the capability of a computer to be used for automated industrial, statistical or architectural design via the use of visual devices.

(2) Computer-Aided Dispatch. A system in which hand written incident cards and conveyor belts in radio dispatching operations are replaced by alphanumeric keyboards (KB,1) and cathode ray tube (CRT) terminals.

CADC Computer-Aided Design Centre. Located near Cambridge in England, this installation's major function is computer aided design (CAD).

CAD/CAM Computer-Aided Design/Computer-Aided Manufacture. A method of production employing designs provided by a computer aided design (CAD) system held in a memory device, to produce finished products by computer or numerical controlled (N/C) machinery.

CAD/CAT Computer-Aided Design/Computer-Aided Test. A method of production employing designs provided by a computer-aided design (CAD) system held in a memory device, to provide final testing of a finished product under the control of a computer. The tests are most often undertaken by automatic test equipment (ATE).

CADMAT Computer-Aided Design Manufacture And Test. A three-year scheme introduced in 1982, under the auspices of the UK Department of Trade and Industry, to promote the greater use of appropriate computer-aided techniques within the electronics industry. CADMAT techniques have been developed to satisfy the requirements of computer-aided design/computer-aided manufacture (CAD/CAM) when applied to the production of computers and electronic equipment, as well as additional specialist requirements for the design, verification layout and inter-connection of circuits, and for manufacture, procurement and testing.

CAFS Content-Addressable File Store. Attachable to a mainframe computer as a peripheral, this stand alone unit permits a very high rate of searching of a structured set of records or database using various sets of fields or keys to identify a particular record.

CAI (1) Computer-Aided Instruction. Enabling a student to converse with a computer, CAI uses a programmed study course that selects topics according to the user's response. Thus students are allowed to progress at their own rate of learning.
(2) Computer-Assisted Instruction. This application of computer instruction provides for a dialogue between a student and a computer program, that informs the user of mistakes as they are made.

CAI/CAL Computer-Aided Instruction/Computer-Aided Learning. The linking together of computer-aided instruction (CAI,1) with computer-aided learning (CAL).

CAIRS Computer-Assisted Information Retrieval System. A computer system that basically provides information to users in response to their requests. The computer has files containing information that can be searched for particular items, in response to the definition of particular user needs.

CAL Computer-Aided Learning. The use of a computer in a system that is employed to aid students in a course of programmed learning.

CAM (1) Computer-Aided Manufacture. A manufacturing production process resulting in finished products, controlled by a computer or numerical controlled (N/C) machinery.
(2) Content-Addressable Memory. A type of memory that can be addressed by its contents or meaning rather than by memory position or location.

CAMAC Computer-Automated Measurement And Control. A standard that defines the physical sizes of modules and racks, as well as defining power supplies, module address lines, data lines and control lines. This in-depth specification (SPEC) is used for the general interfacing of instruments and was originally developed for the nuclear industry to transfer large volumes of data quickly with no errors. *See Figure C.1.*

CAN CANcel. (a) The cancel character instruction is used to indicate data that is in error and should not be accepted. (b) A control code for the deletion of information used in American standard code for information interchange (ASCII).

CANCL CANCeL. This status word is used to indicate that a remote system has deleted information previously transmitted.

CAR (1) Channel Address Register. A storage location that provides information to access a path to a particular register, location or unit of memory where information is stored.

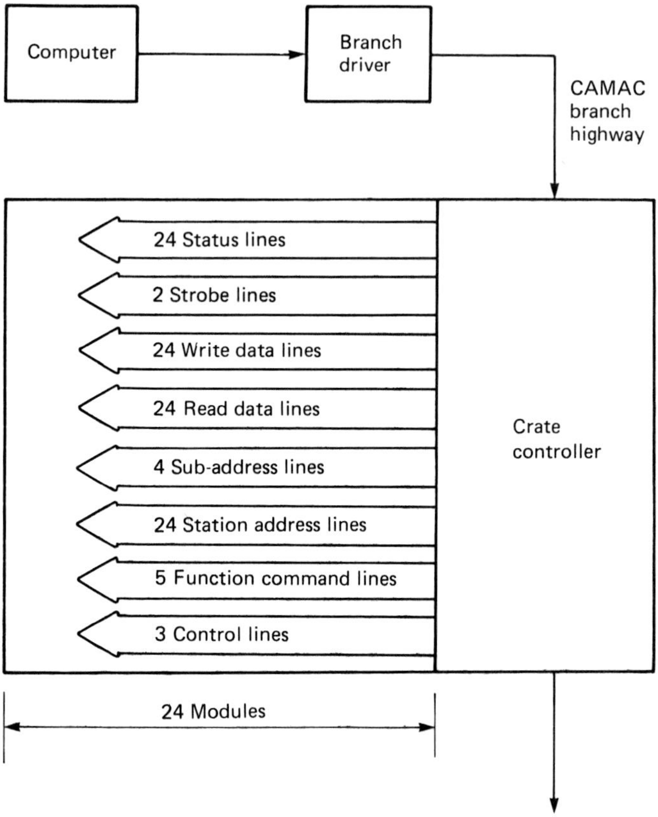

Figure C.1 Configuration of the computer-automated measurement and control (CAMAC) branch highway

(2) Computer-Aided Retrieval. The use of computers in document-retrieval systems (DRS) to replace the simple manual look-up reference systems.

CARR CARRier. An unmodulated radio frequency (RF) source signal of a specific wavelength, usually modulated by frequency modulation (FM) or amplitude modulation (AM) variations, representing the information to be conveyed.

CAS Column Address Strobe. Used in dynamic memory control to find the position of a word in a memory.

CASE Computer-Aided Software Engineering system. Systems that facilitate the implementation of complex automation systems at all stages of an automation project, offering not only standard UNIX (*see* UNIX) tools, but also additional tools for successful project management, such as implementation of databases, and word processing (WP) and project-planning programs.

CAT (1) Computerized Axial Tomography. A computerized X-ray process that produces detailed images of the body's internal organs to be used as a diagnostic tool. The process has a sensitivity 50 times higher than that of ordinary X-ray.
(2) CATalogue. A command in the common algorithmic language (COMAL) computer programming language that permits the directory of all files on a floppy disk (FD (1)) to be displayed on a visual display unit (VDU).

CATV CAble TeleVision. Previously known as 'community antenna television', CATV refers to the use of coaxial cable (COAX,2), or various other cable loops to deliver television (TV(a)) signals to subscribers. Sometimes also known as 'piped aerial'.

CAV CAVity resonance. A cavity formed by a conducting surface provides this effect, its size and shape being chosen to give resonance to a particular frequency of electromagnetic wave.

CAW Channel Address Word. A word that provides an access path to a particular register, location or unit of memory where information is stored.

CB (1) Citizen Band. A frequency band allocated to privately-owned stationary or mobile transmitter/receivers, that are intended for operation by unskilled operators.
(2) Common Base. A particular type of transistor amplifier (AMP,1) that has its base element common to the input and the output (O,2) circuits. *See Figure C.2.*

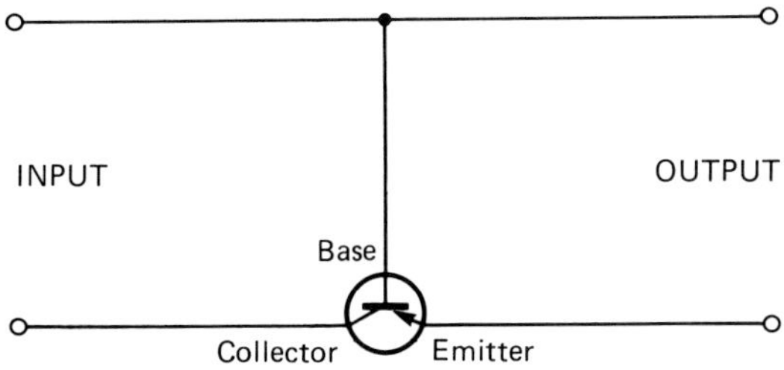

Figure C.2 Basic form of a common-base (CB,2) connection bipolar-transistor amplifying circuit

(3) Central Battery. Employed in a central battery telephone system, in which energy for signalling and speaking is drawn from a power installation at the exchange. Also called a 'common battery', used in a common battery system.

CBI Common Batch Identification. A method by which a number of groups of computer records or documents can be recognized.

CBL Computer-Based Learning. A structured course of teaching and learning that relies on the use of a computer.

CBMS Computer-Based Message System. A term used to describe the communication via electronic mail (E-COM) of keyboarded material which also includes the idea of electronic mailboxes.

CBX Computer-controlled Branch eXchange. Controlled by a computer, this type of telephone exchange is able to offer various additional facilities not normally available on an ordinary non-intelligent telephone exchange.

CC (1) Common Collector. A particular type of transistor amplifier (AMP,1) that has its collector element common to the input and output (O, 2) circuit. *See Figure C.3.*

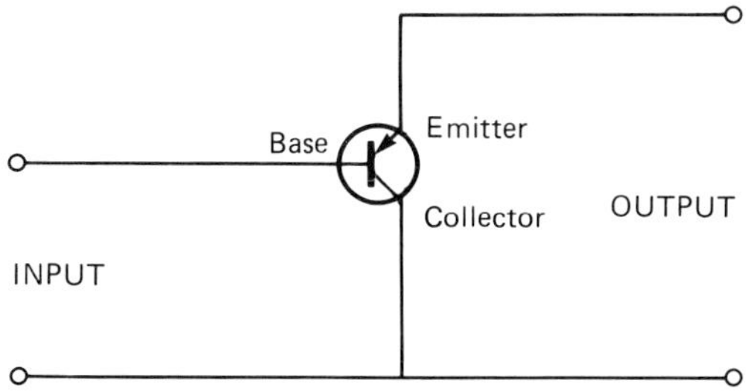

Figure C.3 Basic form of a common-collector (CC,1) connection bipolar-transistor amplifying circuit

(2) CalCulator. (a) A data processor suitable for performing arithmetical operations with little human intervention. (b) A device for carrying out arithmetic, digital and logic operations.
(3) Charge Conveyor. A device or conducting path capable of transferring a flow of electricity.
(4) Chip Carrier. A high-density package for very large scale integration (VLSI) integrated circuits (IC). It may be a leadless chip carrier (LCC), a leadless ceramic chip carrier (LCCC), a pin grid array (PGA) or a leadless land grid array (LLGA). There are many different sizes available.

CCB Channel Control Block. A set of locations in a memory containing information that is accessed by various programs and provides control of the input or output (I/O) of data via a channel (CHNL(b)).

CCCCD (C^4D) Conductivity-Connected Charge-Coupled Device. A hybrid between a charge-coupled device (CCD) and a bucket-brigade device (BBD). It is a CCD that employs doped regions between its potential wells.

CCD Charge-Coupled Device. A unit that is basically a charge-transfer device (CTD), consisting of metal-oxide semiconductor (MOS) capacitors (C, 6), as shown in *Figure C.4.*

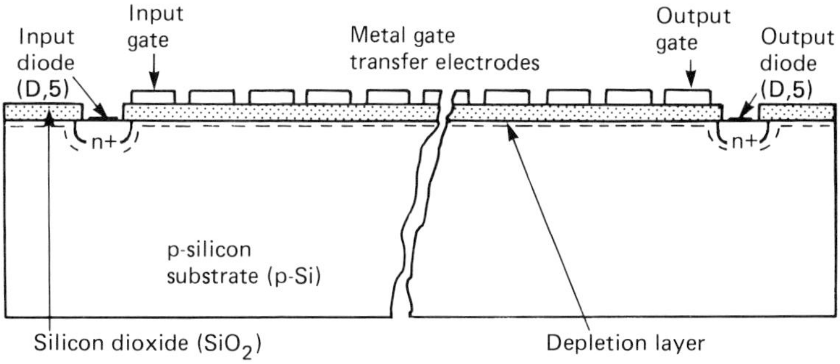

Figure C.4 A cross-sectional view of a charge-coupled device (CCD)

The capacitors are coupled so that charges can be moved through the semiconductor substrate in a controlled manner. These devices can perform a variety of functions including serial memory, signal processing and dynamic filtering.

CCETT Centre Commun d'Études de Television et Telecommunications. A television (TV(a)) and telecommunications research establishment based in France.

CCIR Comité Consultatif International des Radiocommunications. (Consultative Committee on International Radio.) One of the three main organizations within the International Telecommunications Union (ITU), particularly involved in the examination and standards recommendation of long-range radio communications.

CCITT Comité Consultatif International Télégraphique et Téléphonique. (International Telegraph and Telephone Consultative Committee.) One of the three main organizations within the International Telecommunications Union (ITU), concerned with devising and proposing recommendations for international telecommunications.

CCP Console Command Processor. An area in the memory map (MAP, 2) of a microcomputer (MP) using a control program for microcomputers (CP/M) disk operating system (DOS) that allows the user to converse with the operating system (OS) by using the command syntax. Thus the user can accomplish various tasks which include listing a directory on a disk or executing a program file on a disk simply by using the terminal that is tagged on the system.

CCS Common-Channel Signalling. The use of a specific channel (CHNL(a)) for the transmission of information relating to the control of calls in other channels. The channel transfer of network management information.

CCSA Common-Control Switching Arrangement. An American term for a switching facility to corporate tie-line networks. The switching of leased lines in an organization's network is then accomplished by switching equipment. Thus, stations in the network can dial each other without charge, and outside calls can be made by dialling, normally 8 in the USA and 9 in the UK.

CCT CirCuiT. A conducting path comprised of electrical or electronic devices which together perform a desired function.

CCTA Central Computer and Telecommunications Agency. Part of a UK Civil Service department, having responsibility for policy on government administrative telecommunications requirements. It manages the Government Telecommunications Network (GTN) and advises on public purchasing policies in computer and telecommunications fields.

CCTV Closed Circuit TeleVision. A localized non-broadcasting television (TV(a)) system used for selective viewing. Often used as part of a security system.

CCW (1) Counter ClockWise. American terminology for 'anti-clockwise'.
(2) Channel Command Word. A set of binary digits (BITs) treated as a single unit which is transmitted to a channel (CHNL(b)) to indicate the functions that it should perform.

cd CanDela. The unit of measurement of luminous intensity in the Système International d'Unités (SI) system of units. One candela emits 4π of light flux.

Cd CaDmium. A metal having the atomic number 48.

CD (1) Current Density. In a current-carrying medium which can be either a radiation beam or a conductor, CD is the ratio of current to cross-sectional area of the medium.
(2) Compact Disk. An audio disk system in which the disk is only one seventh of the total area of a conventional record or disk, although a single side can play for an hour. Information is carried or encoded as spiral groove of pits and flats in digital (D,2) form. The CD carries a total of over five billion digital sound signal binary digits (BITs) not including additional bits that are used for speed control and error correction tasks.
(3) C or D. (a) Special paper of uniform size and shape which has holes or various marks or arrays that can be sensed by brushes or photoelectric pick ups. (b) A pluggable unit for printed circuit wiring and components.

CDC (1) Code Directing Character. One or more routing indicators, used in telecommunications at the start of a message to determine its destination.
(2) Call Directing Code. An identifying call, usually two letters in length, that is transmitted to an outlying receiver to turn on its printer automatically.
(3) Control Data Corporation. A US-based manufacturer of computers, in particular large processing systems.

CDH Command and Data Handling. Refers to the decoding or storing of an electronic pulse signal or set of signals (command) and basic elements of information (data).

CDI Collector Diffusion Isolation. A form of bipolar process technology.

CDM Colour-Division Multiplexing. In the visible region of the electromagnetic frequency spectrum, CDM is a similar process to frequency-division multiplexing (FDM). Each colour corresponds to a different frequency and a different wavelength.

CDP (1) Checkout Data Processor. A device that has the capability of performing the reduction, summarizing, processing, or input of information and has a set of routines developed to provide a programmer with an evaluation of a program under operating conditions.
(2) Communications Data Processor. A device that has the capability of performing the reduction, summarizing, or input (I/P) and output (O/P), of transmitted data originating from another point.

CDPM Coloured Digital Panel Meter. A meter providing the usual features of conventional panel meters and in addition changes the colour of the display at desired input levels set by trimmer potentiometers (POT). Typically at the lowest input (I/P) the display (D,4) is green, it then changes to yellow, and then to red as the input to the meter is increased.

CD ROM Compact Disk Read-Only Memory. A compact disk (CD,2) that stores information in an identical manner to that used for audio use, but in this case storing computer data. It can offer 500 to 1000 times as much storage as a floppy disk (FD(1)), which is equal to over 500 megabytes (MB) of data or over 120 thousand A4 pages of text.

CDT Control Data Terminal. A point at which items of data that are used to identify, select, execute or modify, may be entered or can leave a communications network.

CDU Central Display Unit. A device (usually a VDU) providing a visual representation at a single location of data obtained from a number of geographical locations.

CE (1) Common-Emitter. A transistor amplifier (AMP) in which

the emitter element is common to the input and the output (O, 2) circuit. *See Figure C.5.*
(2) Customer Engineer. Employed by the original equipment manufacturer (OEM), this engineer is responsible for the maintenance of the system when it is installed at the customer's premises.

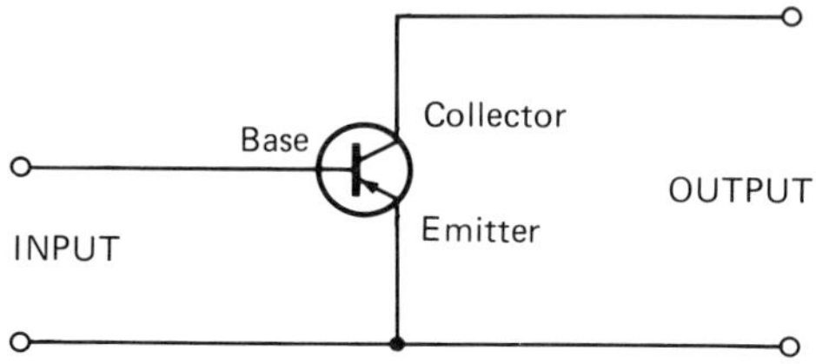

Figure C.5 Basic form of a common-emitter (CE,1) connection bipolar-transistor amplifying circuit

CECC CENELEC Electronic Components Committee. Evolved from the BS 9000 system (BS 9000), the CECC was formed by eleven European nations in 1970 to establish a similar system of quality assurance over a wide range of components under the authority of the European Committee for Electrotechnical Standardisation. The system has the objective of harmonizing specifications and quality assessment procedures for electronic components, by the granting of an internationally recognized Mark and/or Certificate of Conformity. The member countries are Belgium, Denmark, France, West Germany, Ireland, Italy, Netherlands, Norway, Sweden, Switzerland and the United Kingdom. (*See* CENELEC.)

CEE International Commission for the Approval of Electrical Equipment. Comprising the national electrotechnical committees of 22 European countries, the CEE founded in 1946 also includes as observer members Australia, Canada, Hong Kong, Iceland, India, Spain, Japan, South Africa and the USA.

CEEFAX BBC's teletext. A play on words from 'see' and 'facts'. A system which transmits data in the vertical blanking interval of a television (TV) picture. The data can be accessed only by users who have a special decoder unit.

CEGB Central Electricity Generating Board. The CEGB is the British, nationalized agency for the production and distribution of electricity. As such it is one of the largest utilities in the world.

CEN European Committee for Standardisation. CEN comprises the national standards bodies of 14 European Economic Community (EEC) and European Free Trade Association

(EFTA) countries along with Greece and Spain. It publishes European Standards which have to be accepted by a large majority of its members and are published without any variation of text.

CENCER European Committee for Standardisation Certification. The certification body of the European Committee for Standardisation (CEN).

CENELEC European Committee for Electrotechnical Standardisation. The electrotechnical counterpart of the European Committee for Standardisation (CEN). It was formed from a union of CENEL and CENELCOM and comprises the national electrotechnical committees of 15 West European countries as well as over 30 technical committees.

CEOC Colloque Européene des Organisations de Contrôle. Comprising leading independent technical inspection organisations in most European countries, CEOC was formed in 1961. Its aim is to harmonize the practical standards for the inspection of engineering plant and machinery, by the exchange of information.

CEPT Conference Européene des Administration des Postes et des Télécommunications. (Conference of European Postal and Telecommunications Administrations.) A European conference body that makes recommendations for telecommunications practice and standards within Europe.

CERDIP CERamic Dual-In-line Package. A dual-in-line package (DIP) with high reliability characteristics and a very durable and rugged form of housing.

CERN Centre Européene pour la Recherche Nucleaire. Based on the Swiss border of France, this organization is concerned mainly with nuclear physics research.

CF (1) Control Footing. In computing, a minor total that is a summary at the end of a control group or of various control groups.

(2) Central File. May be thought of as a single electronic reservoir or data-deposit, containing a central information file, index and all major application files.

(3) Count Forward. The act of a device adding so that it counts the increases of a common cumulative total of appearances or occurrences of a particular event.

CFD Compact Floppy Disk. A smaller form of floppy disk (FD,1), being 3, 3¼ or 3½ inches rather than the conventional mini disk sizes of 5 and 5¼ inches.

CFDs have a rigid hard plastic shell measuring 80 × 100 × 5 mm (*see Figure C.6*) with a head window shutter on the case that automatically opens and closes when the case is inserted or

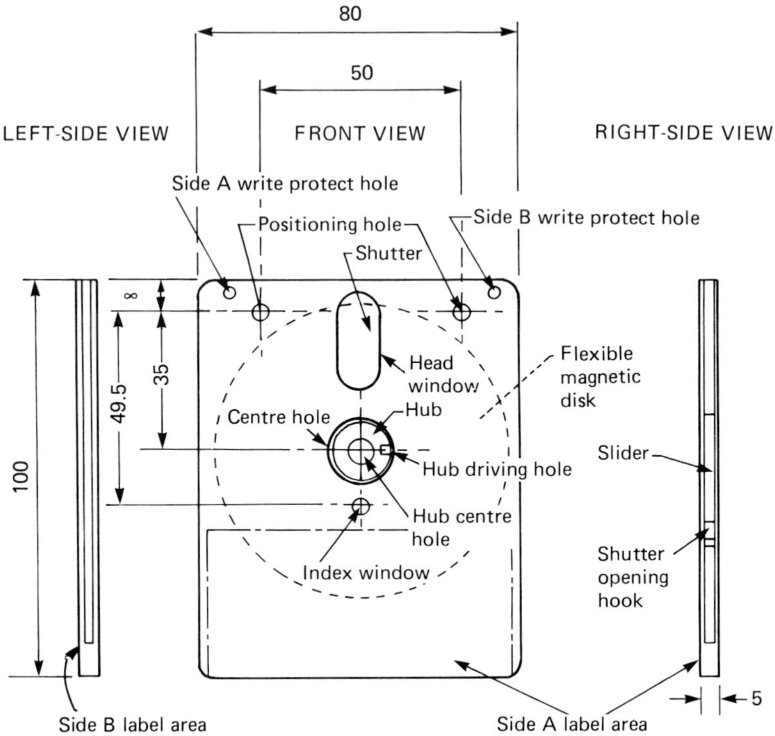

Figure C.6 Component parts and dimensions (in millimetres) of a compact floppy disk (CFD)

pulled out of a compact floppy disk drive (CFDD) system.

CFDD Compact Floppy Disk Drive. CFDDs have been designed to match and drive each size of compact floppy disk (CFD) and are direct, plug-compatible replacements for conventional mini floppy disk (FD,1) drives. The rotation speed, data transfer rate, recording capacity per track and other specifications are exactly similar to those of the mini floppy disks, so existing floppy disk controllers (FDC) can be employed.

CFR Catastrophic Failure Rate. The number of times any type of failure occurs that renders the useful performance of a computer or equipment to zero.

CGB Convert Gray to Binary. The process of changing numerical information from Gray (a type of code wherein, passing from one decimal digit to the next sequential digit, only one binary digit (BIT) changes its value) to Binary (a coding system in which the encoding of any data is done through the use of bits, i.e. 0 or 1).

CGS Centimetre, Gram, Second system. A system of units in which the fundamental units of length, mass and time are the centimetre, gram (g) and second (s) respectively (now generally obsolete).

CH Control Heading. The title or short definition of a control group of records which appears in front of that group.

CHG CHanGe. An alteration to master files or computer programs.

CHILL CCITT HIgh-Level Language. A high-level computer language used particularly for real-time applications such as the control of electronic telephone exchanges, put forward by the Comité Consultatif International Télégraphique et Téléphonique (CCITT).

CHINT CHarge INjection Transistor. A negative resistance field effect transistor (NERFET), in which an extra contact is added to the substrate which collects carriers injected out of the two-dimensional electron gas (2 DEG) over the heterojunction barrier.

CHIP (1) A colloquial expression for capacitors (C,2), resistors (R,2), microprocessors (MP) or other devices manufactured on small pieces (chips) of silicon, ceramic or other materials (i.e. 'chip resistor').
(2) Chip Hermetically In Plastics. Microprocessors (MP) or other devices cut from a larger wafer, usually silicon, hermetically sealed within plastics or resin.

CHKPT CHecKPoinT. A point of time during a machine run where processing is halted momentarily to allow the variables of the run to be recorded on magnetic tape (MAGTAPE).

CHNL CHaNneL (a) A particular path or specific frequency band in communications used for the reception or transmission of signals. (b) In a computer (CMPT) or data-processing system (DPS, 1) a route along which information travels. (c) In data transmission, a means of performing two-way-alternative communication in one preassigned direction. (d) An individual path in a transmission medium, that is separated from other paths, e.g. a multipair cable. (e) A connection between two nodes in a network. (f) In communication theory, the part of a communication system that connects a data source to data acceptance equipment. (g) A complete facility for the transmission, emission or reception of signals, e.g. a television (TV(a)) channel or a radio channel. (h) In a field-effect transistor (FET), the region connecting a current source to the point where charge carriers leave.

CHPS CHaracters Per Second. *See* CPS (1).

CHRG CHaRGe. An excess or deficiency of electrons, which provides a body with negative or positive electrification, i.e. a quantity of unbalanced electricity in a body.

CI Call Indicator. A flagging device in a computer system that shows when branching or transfer of control to a specified closed subroutine has taken place.

C/I Carrier-to-Interface ratio. A specified ratio between a carrier wave on either side of a boundary between two systems or devices.

CICS Customer Information Control System. A system that provides an interface between operating systems (OS), access methods and applications programs to allow remote or local display (D,6) terminal interaction with the data base of the central processor.

CIF (1) Central Index File. (a) A centrally located table containing a set of key characters, treated as a whole, that identify the actual disk records in another permanent disk file. (b) A central file of terms in an automatic retrieval system that is searched by a computer.
(2) Central Integration Facility. A centrally located device whose output (O,2) function is proportional to the integral of the input function with respect to a specified variable.

CIM Computer Input Microfilm. A system whereby images on microfilm are scanned and converted into a digital form and then stored in the memory of a computer.

CIO (1) Central Input Output multiplexor (MUX,2). A centrally-located device that takes low-speed inputs or outputs (O,2) from terminals and combines them into one high-speed data stream, for transmission between an external source and a computer.
(2) Carrier Insertion Oscillator. In single sideband (SSB) transmission, a CIO is used to replace the suppressed sideband at the receiver to permit demodulation.

CIOCS Communication Input/Output Control System. A group of routines which automatically control the performance of input/output (I/O) operations and direct other functions, in a computer system that handles on-line, real-time applications (i.e. a communications system).

CIP Cataloguing In Publication. Data that is held in a National Library's machine-readable cataloguing (MARC) files, providing details on material that is in the process of publication.

CIR CIRcuit. A combination of electrical or electronic devices forming a conducting path, that together carry out a desired function.

CIRC Cross-Interleaved Reed-Solomon Code. An error-correcting code used in the compact disk (CD,2) system.

CISPER Comité International Special de PEtulsation Radioélectriques. An international body formed to set specific limits, test equipment specifications (SPEC), test procedures and electromagnetic interference (EMI) controls throughout most of the world.

CISPR International Special Committee on Radio Interference for the Approval of Electrical Equipment. Formed under the authority of the International Electrotechnical Commission (IEC), the CISPR brings together IEC member committees and other international organizations in the electrical and broadcasting fields.

CIT Call-In-Time. A period of time in which the control of a digital (D,4) computer is transferred from a main routine to a sub-routine. The period has been inserted into a sequence to fulfil a subsidiary purpose.

CIU Computer Interface Unit. A unit that provides a common boundary between automatic data processing systems (ADPS). In communications and data systems it may involve code, format, speed or other changes as required.

CK ChecK. (a) A means of verifying the accuracy of data transmitted, manipulated or stored by any unit or device in a computer system. (b) A process of testing the corrections of machine operations, the existence of certain prescribed conditions within a computer, or the correctness of the results produced by a program.

CKDIG ChecK DIGit. (a) In a character or word, one or more redundant digits (DGT,1) which depend on one another such that if one digit changes, a malfunction can be detected. (b) When carried along with a machine word, one or more digits used in relation to other digits in the word as a self-checking or error-detecting code.

CKO ChecKing Operator. The what-to-do portion of a checking operation.

CKT Circuit. A combination of electrical or electronic devices forming a conducting path, that together carry out a desired function.

CLK CLocK. Used to time events within a piece of equipment or device, a clock provides a timing signal of a precise frequency. Normally crystal-controlled, it is the basic method of producing periodic signals for synchronizing electronic equipment, especially computers.

CM Communications Multiplexor. A communications device that provides for the transmission of two or more messages simultaneously over the same channel (CHNL, (a)). It achieves this by frequency-division multiplexing (FDM), or by time-division multiplexing (TDM).

CMA Circular Mil Area. A unit of area used to indicate a wire size. One circular unit equals the cross-sectional area of a wire one mil (0·001 inches) in diameter.

CMD CoManD. (a) An electronic pulse, signal or set of signals to start, stop or continue an operation. (b) The portion of an instruction word which specifies the operation to be performed.

CML Common Mode Logic. A family of integrated logic circuits, with pairs of transistors coupled by their emitters forming a fundamental part of a circuit.

CMOS or **C/MOS** Complementary Metal-Oxide Semiconductor. A pair of metal-oxide transistors of opposite type used together, as shown in *Figure C.7*, CMOS uses the complementary symmetry of n-channel and p-channel transistors to switch a voltage (V,2) from one circuit to another with very low power dissipation or loss.

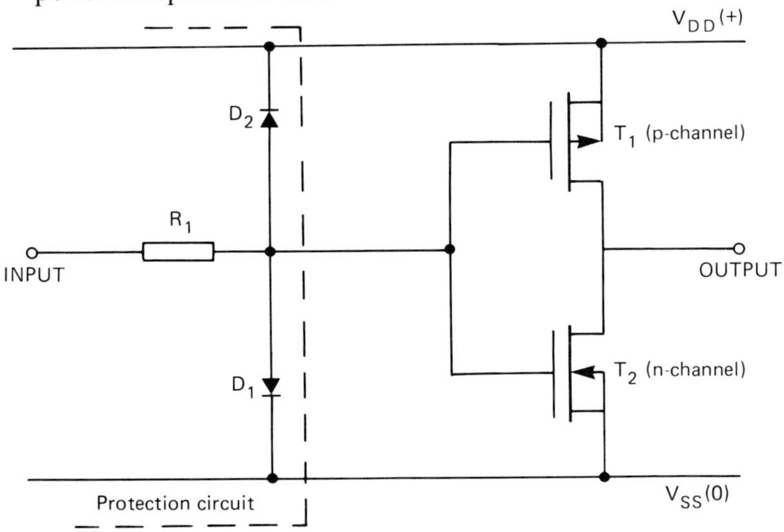

Figure C.7 A complementary metal-oxide semiconductor (CMOS) inverter circuit

CMOSL Complementary Metal-Oxide Semiconductor Logic. A common form of integrated circuit (IC) logic comprising complementary metal-oxide semiconductor (CMOS) gate arrays, characterized by medium-speed operation, very low power dissipation and a wide operating voltage (V,2) range, from +3 to +18 volts (V,1).

CMPLX CoMPLeX. Usually refers to programs or circuits which have increased intricacies of structure or design, which produce significantly faster devices.

CMPT CoMPuTer. An automatic device capable of accepting information, applying prescribed processes to the information, and supplying the results of these processes. The instructions are stored in a memory. The most widely used and most versatile of these devices is the digital (D,4) computer.

CMR (1) Common-Mode Rejection. An ideal differential amplifier (AMP,1), produces an output of zero when the inputs are identical. In practice a small positive or negative signal may occur, known as common-mode rejection.
(2) Communications Moon Relay. A relay station on a moon landing module, used for communication between Earth control and astronauts mobile on the surface of the Moon.

CMRR Common-Mode Rejection Ratio. The measure of the ability of a differential amplifier (AMP,1) to produce a zero output (O,2) for like inputs. *See* CMR,1.

C/N Carrier-to-Noise. Expressed in decibels (dB), C/N is the ratio of the value of a carrier wave (CW,1), to that of unwanted electrical disturbance (noise) after detection.

CNC Computer(ized) Numerical Control. A numerical control system that uses a dedicated stored computer program to perform some or all basic numerical control functions. Usually used to control machine tools such as lathes, etc.

CNCT CoNneCT. The action of using a device that enables electrical or other impulses to be transferred between two pieces of equipment.

CND CoNDition. In the common business-orientated computer language (COBOL), this is an expression of a set of specified minimum (MIN) or maximum (MAX) control values that a specified datum can assume.

CNR Carrier-to-Noise power Ratio. The ratio of the value of a signal wave to that of unwanted electrical disturbance (noise) after detection.

CNT CouNTer. (a) Any electronic circuit, device, sub-assembly or instrument that counts electronic pulses. (b) A device that detects and counts individual particles and photons. (c) A register or storage location in a computer or an electromechanical unit forming part of another device, that is used to accumulate totals.

CNTRL CeNTRaL. Refers to a position in a network or system which is connected with, or is in communication with, all other points in the network or system.

COAM Customer Owned And Maintained. Pieces or sections of equipment that are exclusively maintained by engineers employed by the equipment owners.

COAX (1) COAX. A colloquial expression referring to a coaxial cable.

(2) COAXial cable. A cable consisting of concentric outer and inner conductors, separated from each other by insulation.

COBOL Common Business-Orientated Language. A computer language designed for commercial data processing (DP,1) such as accounting and financial work. It is a specific language by which business-data processing procedures may be precisely described in standard form.

CODASYL COnference for DAta SYstems Languages. Created by the US Department of Defense along with computer users and manufacturers, this group has specified a number of application-independent and manufacturer-independent software tools (e.g. COBOL).

CODEC COder-DECoder. An analogue-to-digital (A/D,2) and digital-to-analogue (D/A,2) converter. It is used to convert analogue signals such as speech, music or television (TV(a)) to digital (D,4) form for transmission over a digital medium and back again to the original analogue form. *See Figure C.8.*

CODIC COmputer-DIrected Communications. Transfer of in-

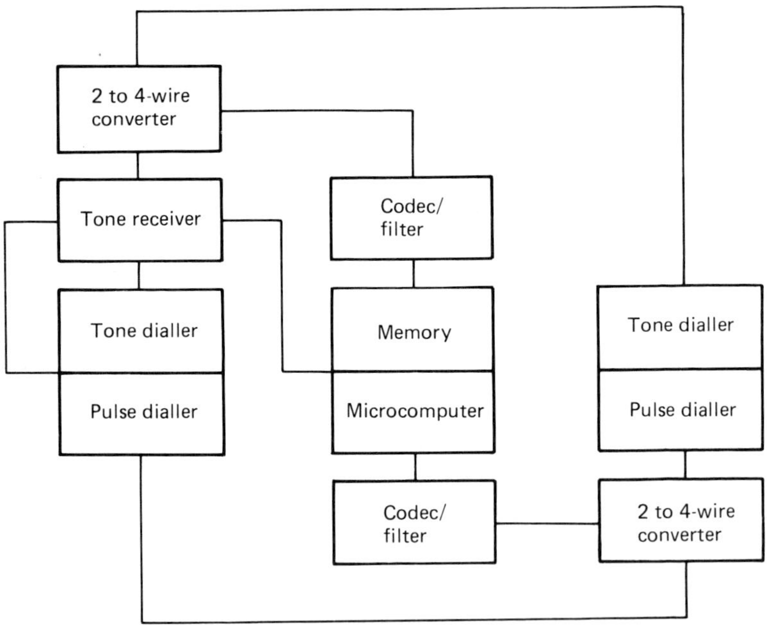

Figure C.8 A typical design of a voice channel telecommunications coder-decoder (CODEC)

formation in various forms, routed via a computer from one point, person or device to another.

COED Computer-Operated Electronics Display. A form of display (D,6) at large sports stadiums or on advertisements consisting of lights used to construct words or shapes, controlled by computer.

COGO Coordinate GeOmetry. A computer language that is used for solving coordinate geometry problems in civil engineering.

COHO COHerent Oscillator. A device that produces radiation waves having pure well-defined oscillations, such as in a laser (LASER).

COL Computer-Orientated Language. A programming language requiring little translation. This type of program usually runs efficiently on a related computer, but requires extensive translation for use on another variety of computer.

COM Computer Output Microfilm. A microfilm printer that will substitute directly for a line printer or tape output (O,2) by taking output directly from a computer. It may also be referred to as Computer Output Microfilmer, Computer Output Microform, Computer Onto Microfilm, Computer Output on Microfilm, Computer Output Microfiche or Computer Onto Microfiche.

COMAL COmmon Algorithmic Language. A conventional programming language that was originally devised to meet a specific educational need. It was devised and developed in Denmark by the Danish lecturer Borge Christensen, who was dissatisfied with beginners all-purpose symbolic instruction code (BASIC) as a vehicle for introducing programming to his students.

COMFET COnductivity-Modulated Field-Effect Transistor. Operating in much the same way as metal-oxide silicon field-effect transistor (MOSFET), the COMFET, however, requires a smaller chip (CHIP), approximately one-tenth of the size of a comparable conventional MOSFET.

COML COMmercial Language. Generally procedural, this type of computer language is designed for data processing (DP,1), accounting and financial work (*see* COBOL).

COMM COMMunication. The process of transferring information in various forms from one point, person or device to another.

COMM-STOR COMMunications STORage unit. A single- or dual-drive diskette system used for storing messages received on line or other locally-prepared information ready for transmission.

COMPAC Computer Output Microfilm PACkage. A package that substitutes directly for a line printer or tape output (O,2), being a microfilm printer that will accept output information directly from a computer.

COMPEDEX COMPuter Engineering inDEX. A computer file containing information covering all branches of engineering.

COMPUNICATIONS COMPUters and commuNICATIONS. This contraction is a colloquial term for the joint use of computers and communication systems.

COMSAT (1) COMmunications SATellite. An artificial unmanned satellite in Earth orbit that provides communication links between locations on Earth which can be widely separated. It accepts and relays suitably modulated signals between earth stations.

(2) COMmunications SATellite Corporation. A privately owned American corporation established by statute as the exclusive international satellite carrier and representing the USA in the International Telecommunications Satellite Consortium (INTELSTAT).

COMTEC COmputer Micrographic and TEChnology. Based in California, this is an association of manufacturers and users of computer output microfilm (COM) equipment.

CONVT CONVerT. (a) To transfer information from one number base to another. (b) To alter numerical information from one number base to another.

COPANT Pan-American Standards Commission. Comprising the national Standards bodies of the USA and eleven Latin American countries, COPANT is a co-ordinating organization concerned with the regional implementation of International Organisation for Standardisation (ISO) and International Electrotechnical Commission (IEC) standards and recommendations.

CoPs COmponent Placement System. A technique that supplies chip capacitors (C,6) in a reeled format so that users may employ automatic placement. The capacitors are packaged in a continuous cardboard carrier strip, sandwiched between two layers of transparent tape with Electronic Industries Association (EIA) standard sprocket-drive holes.

COS COSine (a) A mathematical term which is the sine complement of a given angle or quantity. *See* SIN (a).

(b) A standard Pascal function used to plot a cosine (COS (a)) curve.

COSMOS A synonym for CMOS or C/MOS.

CP (1) Clock Pulse. Used to control the timing of count characters, read circuits or other related functions, this pulse is located near recorded data.

(2) Critical Path. The overall time to complete a project cannot be less than that needed along the critical path. Any delay along this path can delay the whole project, whereas small interruptions along non-critical paths do not.

CPA Critical Path Analysis. *See* critical path (CP,2), critical and program evaluation and review technique (PERT).

CPC Computer Power Centre. A power device that requires only one connection to the mains power, providing regulated power that can be distributed to various machines in a computer room. It also provides a computer with regulated power and protection from power disturbances.

CPE Central Processing Element. An element that can provide a complete 2, 4 or 8 binary digit (BIT) slice across the data processing (DP,1) section of a computer. It enables functions to be executed in one, rather than several, microcycles.

CPI Characters Per Inch. A measure of the number of individual characters (including necessary spaces) that a printer, teletypewriter (TTY) or similar device can print along a line.

CPL Combined Programming Language. A very high-level language developed jointly in the 1960s by the University of Cambridge and the Institute of Computer Science.

CPM (1) Cards Per Minute. A measure of the rate at which a card reader is able to sense and read information punched on cards.

(2) Critical Path Method. Used for scheduling and controlling large projects, this technique of management is particularly useful for projects that involve a number of phases which depend upon one another. *See* CP,2, critical path.

CP/M Control Program for Microcomputers. An operating system (OS) written to run on 8080/8085/Z80 based microcomputers. It is the most widely used disk operating system (DOS) and a large number and variety of languages and utility programs have been written to run under CP/M. There are several versions available.

CPS (1) Characters Per Second. (a) The number of individual letters, numerals or other symbols which are printed in one second. Usually this abbreviation occurs when discussing the speed of characters-at-a-time printers. (b) A measure of the rate of transmission normally between a computer and terminal devices. As various numbers of binary digits (BITs) are required to represent a character, there is no simple correlation between bits per second (BPS) and CPS.

(2) Cycles Per Second. Now superseded by hertz (Hz), CPS is the unit of frequency equal to the number of complete waves or

cyclic processes that occur every second (s).

CPT Colour Picture Tube. A cathode-ray tube (CRT) designed to produce coloured pictures for television (TV(c)). The image depends on the excitation of different phosphors to produce three primary colours, red, green and blue. The basic construction of a CPT is shown in *Figure C.9*.

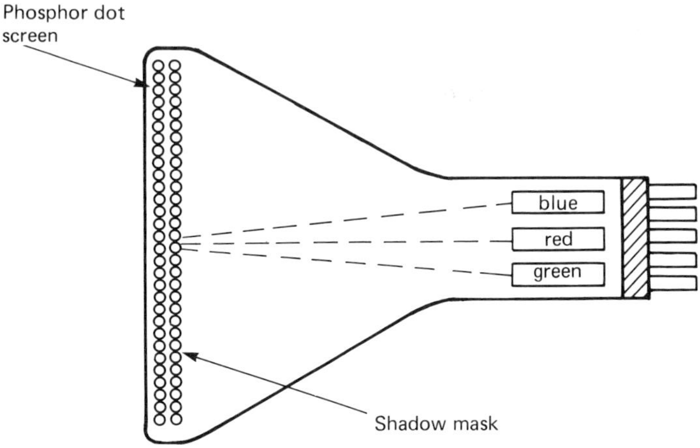

Figure C.9 The basic construction of a colour-picture tube (CPT)

CPU Central Processor Unit. The portion of a computer that controls the operation of the whole system, where instructions are executed and computations performed. It contains the main storage, arithmetic unit and special register groups. Sometimes also known as Control Processor Unit.

CPY CoPY. To reproduce data in a new location, replacing the information that was previously stored there without altering the source of the original data.

CR (1) Carriage Return. (a) The action of moving a carriage back to the beginning of a line on a printer or teletypewriter (TTY). (b) A key on many computer systems or teletypewriters (TTY), the operation of which returns the carriage or transmits a carriage return character, from which the computer transmits a line feed character which in turn generates a new line.

(2) SYNCHRO Receiver. This device has the same electrical connections as a synchro transmitter (CX). The stator-stator leads, however, are connected back to back with the transmitter while the rotor leads are in parallel. This enables the receiver (CR) to follow the transmitter (CX).

CRC Cyclic Redundancy Check. A method of checking a transmitted bit stream for errors. The receiving station compares the transmitted value of information on blocks of data with its own computed value and if they are equal then no error has occurred.

CRD Constant Ringing Drop. A relay employed in burglar alarm systems which, when activated even momentarily, will remain in a closed or alarm state until reset. Often a key is required to reset the relay and turn off the alarm.

CRJE Conversational Remote Job Entry. Enabling users interactively to edit files, this method of running a computer does not allow them to interact with tasks submitted to lower priority processes.

CRN Cellular Radio Network. Cells based on an idealized hexagonal grid which employs the principle of frequency reuse, *see Figure C.10.* At the centre of each cell is a radio base station which has been assigned a set of radio channels (CHNL(a)) for communication and signalling purposes. Groups of adjacent cells may share all the available radio channels. Such groups are called clusters.

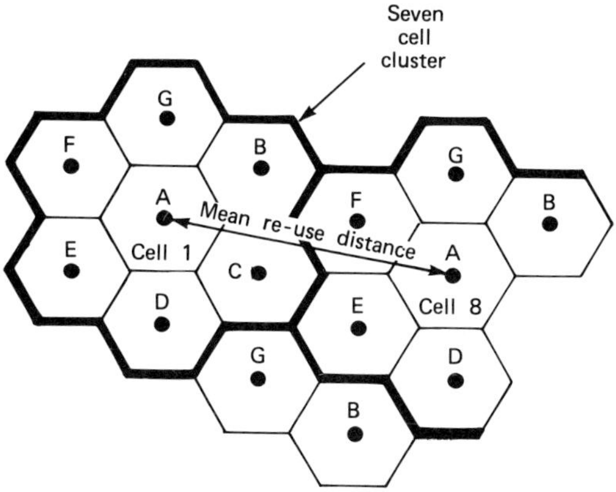

Figure C.10 A cellular radio network (CRN) showing a seven cell cluster and its relationship to other cells. Cells with similar letters use the same frequency.

CRO Cathode-Ray Oscilloscope. A device that permits the visual inspection of high-frequency waveforms which would be totally impossible to observe by any other means. Most CROs incorporate a cathode-ray tube (CRT), but a few employ a flat liquid crystal display (LCD).

CROM Control Read Only Memory. Designed and microprogrammed to decode control logic, this special type of read-only memory (ROM) is an important component in several types of two-chip microprocessors (MP); one chip contains the CPU and the other a CROM.

CRR Constant Ringing Relay. A relay used in burglar alarm systems which, when activated even momentarily, will remain in a closed or alarm state until reset. Often a key is required to reset the relay and turn off the alarm.

CRT Cathode-Ray Tube. A vacuum tube with a viewing screen as an integral part of its envelope, it converts electrical signals into a visible form. CRTs consist of an electron gun that produces an electron beam, a grid that alters the brightness by varying the beam intensity and a luminescent screen for the display (D,6). The electron beam is moved across the screen either by deflection plates or magnets. Focusing of the beam may also be achieved by electrostatic or electromagnetic means or by a combination of methods.

CRTC Cathode-Ray Tube Controller. A circuit or device employed to provide control for a cathode-ray tube (CRT).

C/S Cycles per Second. Now superseded by hertz (Hz), C/S is the unit of frequency equal to the number of complete waves or cyclic processes that occur every second (s).

CSA Canadian Standards Association. A body formed to set safety standards and qualify equipment.

CSCR Complementary Silicon-Controlled Rectifier. A specific type of silicon-controlled rectifier (SCR) that controls the current flows from anode to gate instead of from gate to cathode.

CSECT Control SECTion. A main sequence of instructions or data in a computer program, that can be transferred, deleted or replaced with a control section from other program segments.

CSIPR Comité Special et International sur les Parasites Radiotélégraphiques. Set up by the International Electrotechnical Commission (IEC) this committee establishes standards for telecommunications equipment, in particular the control of radio frequency interference (RFI).

CSIRO Commonwealth Scientific and Industrial Research Organisation. A research organization based in Australia.

CSL Computer-Sensitive Language. A programming language written so that it is totally or partially dependent on the type of machine used.

CSMA Carrier-Sense Multiple Access. A formalized method by which communication is carried out in broadband networks, including local area networks (LAN), such as Ethernet.

CSMA-CD Carrier-Sense Multiple Access-Collision Detection. A particular formalized method by which communication is carried out in broadband networks, including local area networks (LAN), such as Ethernet.

CSMP Continuous Systems Modelling Program. A computer

program that provides a digitally produced simulated analogue system.

CSO Colour Separation Overlay. A superimposition technique used in colour television (TV(a)). A colour area such as blue is selected and viewed by one camera, and the output of another camera is superimposed into the blue area of the first. The complete picture is then transmitted.

CSP Chloro-Sulphonated Polyethylene. A material used for cable sheathing which is heat resistant, oil resistant and flame retardant (HOFR).

CSR (Controlled Silicon Rectifier) Thyristor. 'CSR' represents 'thyristor' on parts lists and annotated schematics. *See* SCR.

CSW Channel Status Word. An individual element of a table that indicates the status of input-output (I/O) channels.

CT (1) Cordless Telephone. Basically a telephone in which the cord between the handset and base unit has been replaced by a fixed frequency radio communication. In spite of its extra complexity it is nevertheless a much simpler device than a digital cellular radio.

(2) Control Transformer. This has the same construction as the synchro transmitter (CX). In this the transmitter and a control transformer are connected together in such a way that a voltage (V,2) is induced in the rotor of the control transformer, dependent on the angle between the two rotors. When both are at the same angle the voltage will be at maximum; when they are at 90° the voltage will be zero. However, in practice the output shaft and input shaft are deemed to be aligned when the output voltage is zero, thus making the design of a servo system more simple.

CTC Conditional Transfer of Control. A machine instruction that can cause a departure from the normal sequence of executing instructions in a computer.

CTD Charge-Transfer Device. A semiconductor device that is capable of transferring discrete packets of charge from one position to the next. It can be used for the storage of charge over a short time, the time being dependent on the material the device is made from.

CTI Comparative Tracking Index. The maximum (MAX) voltage (V,2) at which a material can withstand 50 drops of a test solution, normally 0.1% by mass of ammonium chloride in distilled water, without any tracking occurring. The test voltages and therefore CTI should be divisible by 25.

CTL Complementary Transistor Logic. The use together of a pair of transistors, in a logic circuit, of opposite type. For example a pair of n-p-n and p-n-p transistors, manufactured on the same integrated circuit (IC). Such elements are often constructed from complementary metal-oxide semiconductor (CMOS) field-effect transistors (FET). *See* n-p-n and p-n-p.

CTNE Compania Telefonica National de España. The Spanish post, telegraph and telephone (PTT) administration authority.

CTS Clear To Stand. This control line runs to a terminal from a modulator/demodulator (MODEM) indicating that a carrier is present and that data may be transmitted. An RS-232 standard (*see* RS-232C).

CUG Closed User Group. A collection of computer system users who can treat one system as dedicated to their own particular purposes, having no access to facilities outside of those available to the group, and not permitting access to users who are not members of the CUG.

CUTS Cassette User Tape System. A system, normally microcomputer-based, that employs any of a number of types of digital cassette recorders to record information from any user specified instrument, in formats that can be read by most types of terminal.

CV A video tape format developed by Sony in Japan.

CVD Chemical Vapour Deposition. A process for making optical fibres (OF) where silicon along with other glass-forming oxides and dopants is deposited at high temperature onto the inner wall of a fused silica tube. From this tube a long thin fibre is finally drawn.

CVPO Chemical Vapour Phase Oxidization. A process for the production of graded index (GI) optical fibre (OF), which have a loss of less than 10 dB/km and a bandwidth greater than 300 MHz-km.

CVT Constant Voltage Transformer. A transformer (XFMR) that can provide a substantially constant value of voltage (V,2) independently of the current (I) supplied to it.

CW (1) Carrier Wave. The basis frequency of an unmodulated signal, being sinusoidal oscillation or electromagnetic wave. This wave is then varied or modulated in a desired manner. The term can also be used to describe a direct current (DC,1) which is intended for modulation.
(2) Continuous Wave. A transmitted radio or radar wave that maintains both constant amplitude and constant frequency.

CWP Communicating Word Processor. A specialized desktop computer having copying and communications facilities. It can send information to and receive pages of text in seconds from another word processor by means of a network which allows rapid inter-office or site-to-site text communications.

CX SYNCHRO TRANSMITTER. The simplest form of the

classic synchro position measuring device. It consists of a moveable rotor winding and three symmetrical fixed stator windings. *See Figure C.11.*

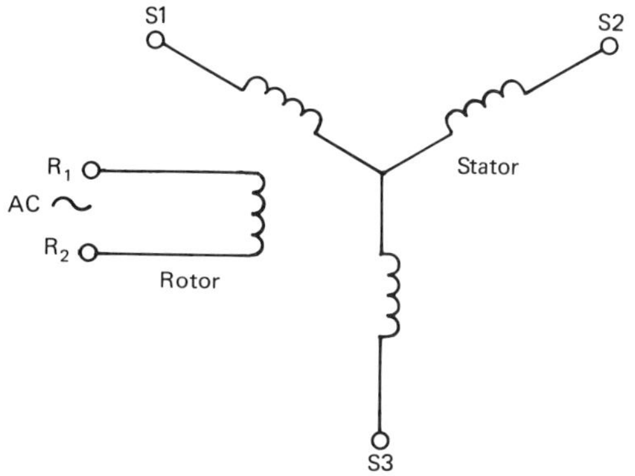

Figure C.11 A synchro transmitter (CX)

D

d (1) Deci. A decimal prefix commonly used in association with a base unit in the Système International d'Unités (SI) system of units indicating a multiplication factor of 10^{-1} of that unit. For example, one decibel (dB) is equal to 0.1 bels.
Although this particular prefix is not one of the multiples that increase in steps of 10^3 preferred by SI, it is used and limited to occasions where the recommended prefixes are inconvenient.
(2) Drain of a field-effect transistor (FET). The connection to the channel (CHNL) which majority carriers pass on leaving the channel. Corresponds to collector (c,2) in bipolar transistors.

D (1) One of the letter designations given to the most commonly used sizes of 1.5-volt (V,1) batteries. Originally a USA nomenclature this type of letter description is gaining increasing popularity over the International Electrotechnical Commission (IEC) letter/number system as it is able to simplify the selection of batteries. *See* battery-size cross-reference guide Appendix.
(2) The hexadecimal (HEX) symbol for binary 1101, decimal 13.
$D_{16} = 13_{10} = 15_8 = 1101_2$.
(3) Density. (a) In printed-circuit-board (PCB) design or circuit design in general, the physical closeness of components or devices. (b) Applied to computers, the closeness of space distribution on storage media such as magnetic drums, magnetic tapes (MAGTAPE), or cathode-ray tubes (CRT).

(4) Digital. The use of particular whole numbers to represent quantities in calculations; on/off, open/closed and true/false are all dual-state conditions that are capable of expressing in digital form all information stored, transferred or processed.
(5) Diode. An electronic device with two electrodes, an anode and a cathode. Diodes are most commonly used as rectifiers. The symbol D is used to represent diode on parts lists and to annotate schematic diagrams.
(6) Display. A visible representation of information which can be produced through various devices such as cathode-ray tubes (CRTs), plasma and liquid crystal displays (PDs and LCDs) and light emitting diodes (LEDs).

da DecA. A decimal prefix used in association with a base unit in the Système International d'Unités (SI) system of units indicating a multiplication factor of 10^1 of that unit. For example, one decametre is equal to 10 metres.
Although this particular prefix is not one of the multiples that increase in steps of 10^3 preferred by SI, it is used and limited to occasions where recommended prefixes are inconvenient.

DA (1) Data Acquisition. The process undertaken by a data acquisition system (DAS), where data is collected at one computer or system location from various other locations. It is then converted into digital (D,4) information for operating, supervisory or accounting records.
(2) Direct Access. (a) Refers to the access or storage of data with reference to its own location and not dependent upon a previously accessed position. (b) A memory device that allows a memory location to be accessed independently of its position, and information in various stores to be accessed in the same amount of time.
(3) Differential Analyser. A device used for solving differential equations, usually an analogue computer, which employs interconnected integrators to solve the equations.

D/A (1) Digital-to-Analogue. The conversion of digital information and representation used in computers to the analogue signals necessary to drive speakers, meters and various other devices. The conversion is employed so that computers can talk, and operate with the outside world.
(2) Digital-to-Analogue converter. A unit which provides continuous analogue output (O,2) channels (CHNL(h)) from digital (D,2) input signals. *See Figure D.1.* Typically the outputs can range from 0 to +10 volts (V) and can be used as the control voltage for servo-mechanisms.

DAA Data-Access Arrangement. A small wall-mounted enclosure which contains an isolating transformer, that has been

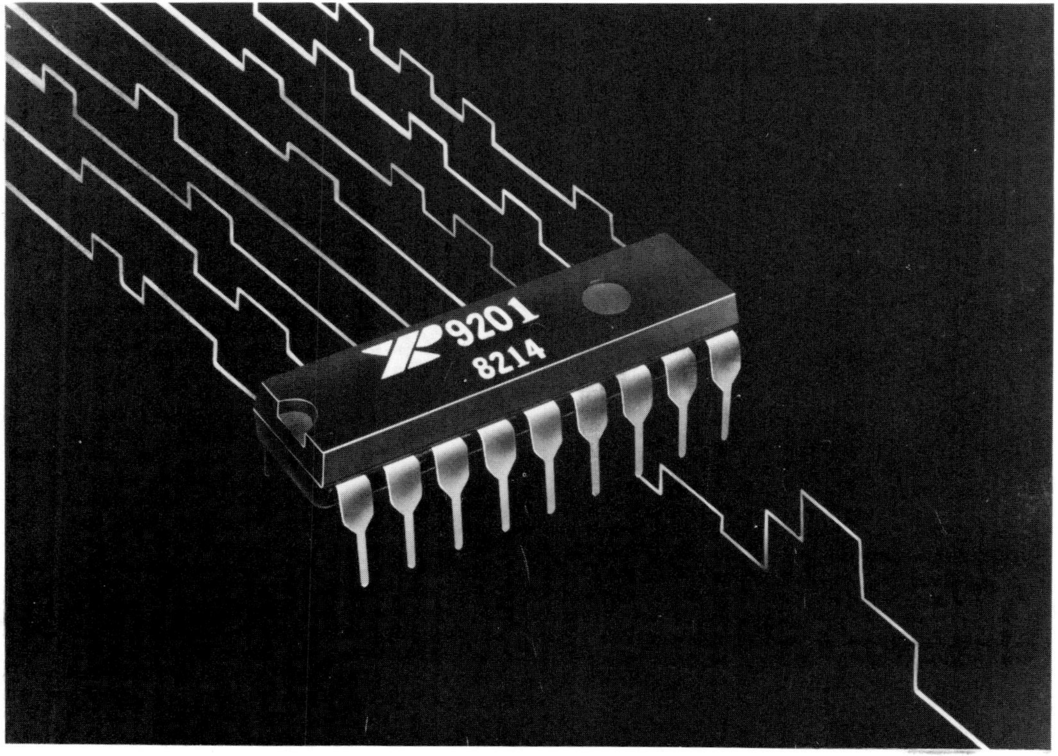

Figure D.1 An 8-bit digital-to-analogue converter (D/A,2) (*courtesy* Thame Components Ltd)

designed to prevent signals harmful to a direct-dial telephone network from being sent down the network lines. A DAA is required in the USA to protect the network from some customer-owned modulator demodulators (MODEMs). Also known as a 'direct-access arrangement'.

DABS D ABSolute. An operator in the FORTH computer programming language to replace the top of a stack with the absolute value of 'd'.

DAC (1) Digital-to-Analogue Conversion. The process of converting a digital signal to an analogue form. (Also D/AC.)
(2) Digital-to-Analogue Converter. A unit that provides analogue output (O,2) channels (CHNL,1) from digital input signals. Typically the output can range from 0 to +10 volts (V) and can be used as the control voltage for servo-mechanisms.
(3) Data Acquisition and Control System. A wide variety of real-time (RT,1) applications, process control, and high-speed data acquisition (DA) can be handled by a DAC. Individually designed, each system is tailored to meet specific requirements.

DACOM DAtascope Computer Output Microfilmer. An early output (O,2) device used for computer output microfilm (COM) devices.

DACS Data Acquisition and Control System. *See* DAC, 3.

DAM Direct-Access Memory. A particular type of memory device that permits the access of a memory location independently of its position. The items stored can therefore be accessed in the same amount of time whatever their position.

DAP Distributed Array Processor. Attached as a peripheral to a mainframe computer, this device consists of a matrix of a large number of small processors that permit extensive calculations to be performed.

DAR (1) Damage Assessment Routine. A component of a computer operating system (OS) that endeavours to make a recovery from a software failure.
(2) Daily Activity Report. A report generated by computer concerning the daily uses and operation of programs, routines and subroutines.

DARPA Defense Advanced Research Projects Agency. An organization in the USA which can provide support for the development of military and industrial projects such as rapid optical ocean surveillance testbed (ROOST) equipment.

DAS Data-Acquisition System. A system that collects or acquires data at one computer or system location from various other locations and then converts it into digital (D,4) information.

DASD Direct-Access Storage Device. A device that allows

access time for any stored information on the particular medium to be independent of its location. Generally primary computer memory is direct, disks have pseudo-direct access and tapes have sequential access.

DASM Direct-Access Storage Media. Capable of storing information and computer programs, these media allow rapid access to individual elements. The access is independent of the location of the element required and of the location of the last element of data accessed.

DATACOM DATA COMmunications. (a) A general term used for devices or systems that transmit data from one point to another. (b) A communications network employed by the US Air Force covering the whole world. (c) A data communications service covering and linking more than 60 cities in the USA, set up by Western Union.

DAV (1) DAta Valid. Recognition given in a computer data entry system indicating that data entered has passed extensive arrays of data validation and error checking features.
(2) Data Above Voice. A system of digital (D,4) data transmission that carries information on a frequency above that used for voice transmission. *See* DUV.

dB DeciBel. A unit of measurement of the relative strength of a signal parameter, e.g. power or voltage (V,2).

DBAM Data-Base Access Method. A generic computing term used to describe various processes such as direct access (DA,2) and random access.

DBI Data-Base Index. An index used by the System Development Corporation (SDC), that provides information to all the files of computer information held by the organization. It is employed to help in the selection of a suitable database for a user requirements.

dBM DeciBel Milliwatt. Employed in communication work, decibel (dB) referenced to one milliwatt is a measure of absolute power. Zero dBM is equal to one milliwatt.

DBMS Data-Base Management System. The primary control software used in manipulating computer information files.

DBOS Disk-Based Operating System. Software contained in a computer which allows it to control the operation of all activities related to the use of magnetic disks in the system.

dBrn DeciBels above Reference Noise. Expressed as a ratio, dBrn is a comparison of the decibels (dB) above a certain reference noise to that reference noise. A reference noise is the magnitude of circuit noise that will produce a circuit noise meter reading equal to that produced by 10–12 watts (W,2) of electric power at 1000 hertz (Hz).

DBS Direct Broadcast Satellites or Direct Broadcast by Satellite. Satellites, normally in geostationary orbit, that operate at relatively high output levels to provide television (TV) or other services to users with very small dish aerials. Transmission from DBSs is often scrambled so only subscribers with the correct decoding equipment will be capable of receiving programmes.

dBW DeciBel Watt. Employed in radiated power transmission, it is decibel (dB) relative to one watt (W,2).

DC (1) Direct Current. Having a unidirectional flow this form of the movement of electric charge normally has a constant direction and a substantially constant magnitude. An example of direct current is that produced by a storage battery or cell.

(2) Data Collection. The process whereby data from one or more points is brought to a central position.

(3) Data Conversion. Essentially the process achieved by a hybrid or monolithic device that converts analogue data into digital (D,4) format or vice versa.

(4) Device Control. A character that does not form part of any message or information, this element of a character code is employed to effect control of the operation of equipment, by switching devices on or off. Usually used in telecommunication systems, but also sometimes used with other data-processing equipment.

(5) Direct Coupled. A particular design of circuit connected by resistance or inductance in such a manner as to permit direct current (DC, 1) to pass through the coupling.

(6) Data Channel. A data path which allows two-way communication between input/output (I/O) devices and a main memory in a computer.

(7) Double-Crucible. A process which uses two concentric crucibles to produce optical fibres (OF), as shown in *Figure D.2*: one for the glass forming the core and another for the glass forming the cladding. The complete fibre, including the cladding, is drawn from the bottom. Diffusion of dopant to each of the glasses produces a graded index (GI) fibre.

DCA Defence Communications Agency. A US Department of Defense agency which now controls the packet-switched advanced research projects agency network (ARPANET).

DCB Data Control Block. A storage area holding control or status information employed by file management routines when storing and retrieving data.

DCCU Data Communications Control Unit. Messages that are received by a central terminal unit are scanned by the DCCU and transferred to the central processor unit (CPU).

DC-DC Direct Current to Direct Current. A DC-DC converter

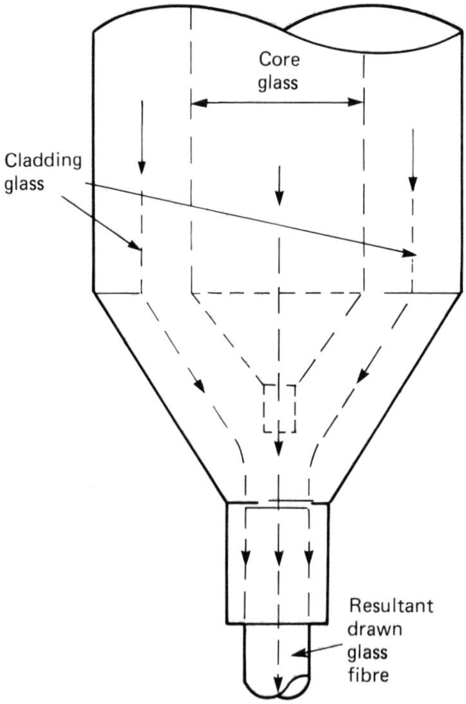

Figure D.2 The double-crucible (DC,7) process for producing drawn optical fibres (OF)

or regulator is a unit that is itself supplied from a direct current (DC,1) source providing a DC output (O,2). It can be employed in systems where a raw DC is available, giving a more stable DC output, plus isolation between input and output.

DCE (1) Data Communications Equipment. Usually used in reference to a modulator demodulator (MODEM), this term describes any piece of equipment that interfaces with a data communications network.

(2) Data-Circuit terminating Equipment. An active unit in a data station or terminal that connects a data transmission line to the terminal equipment. The interface unit controls all connections and provides signal or code conversions if required.

DCMT DeCreMenT. The progressive lessening or decay of a free oscillation due to the expenditure of energy. It is the ratio of two successive values of oscillation of the same sign.

DCTL Direct-Coupled Transistor Logic. A particular type of logic circuit that employs only direct-coupled (DC,5) transistors as active elements. *See Figure D.3.*

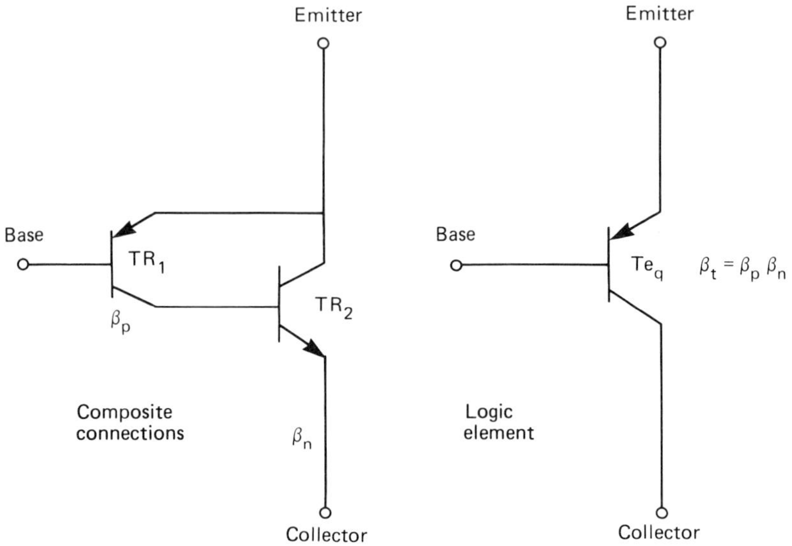

Figure D.3 The composite connection of complementary transistors, forming a direct-coupled transistor logic (DCTL) element

DCUTL Direct-Coupled Unipolar Transistor Logic. A direct-coupled transistor logic (DCTL) circuit that employs field-effect transistors (FETs) as active elements (unipolar transistor being an obsolete term for field-effect transistor).

DD (1) Digital Display. (a) A display (D,6) of digital (D,4) orientated numerals. (b) Generally on a cathode-ray tube (CRT), a visual display of numeric, alphabetic or graphic material. (c) Information that is provided in a digital (D,4) form. (d) A display (D,6) of data in a decimal form.
(2) Data Definition. Commonly used in the job control language (JCL), where it is employed to connect external input/output (I/O) devices to an internal named file or data set (DS) associated with a particular step used in a computer program.

DDA Digital Differential Analyser. (a) An incremental computer that employs a digital (D,4) integrator for computing. (b) A differential analyser that employs digital (D,4) representations for all analogue quantities used. (c) Having specific circuits designed to carry out mathematical integration, a DDA is part of a digital (D,4) computer.

DDC Direct Digital Control. An automated action in a computer control system whereby the signal to a final control element is obtained by a digital (D,2) device. Such elements are used to perform process control tasks for production systems.

DDCE Digital Data Conversion Equipment. Equipment employed in the conversion of digital data (DD,1(c)) into a form other than digital.

DDCMP Digital Data Communication Message Protocol. Governing parallel, serial-synchronous, or serial-asynchronous data transmission and reception, this uniform discipline is used for the transmission of data between stations in a multipoint or point-to-point data-communication system.

DDD Direct Distance Dialling. Information conveyed that automatically establishes, controls, supervises and releases a long-distance telephone call. It enables the telephone user to call outside his local area without operator assistance. In the UK and some other countries, this is called subscriber trunk dialling (STD).

DDI Direct Dialling-In. Allows a telephone caller direct access to any particular public automatic business exchange (PABX) extension without involving the exchange operator. Also known as 'direct inward dialling' (DID).

DDS (1) Dataphone Digital Service. An American communications service provided by the Bell System which eliminates the need for modulator demodulators (MODEMs) by transmitting data in a digital (D,4) form rather than an analogue one.
(2) Dewey Decimal System. A particular form of decimal classification system, developed by Melvil Dewey. It is one of the most widely used systems for cataloguing documents.

DDT (1) Digital Data Transmitter. A package of computer software designed to assist in the isolation and correction of errors in a routine or program.
(2) Dynamic Debugging Technique. A computer technique that makes use of a debug program, especially designed for continuous debugging of user-written programs.

DE Decision Element. In a computer, a DE circuit undertakes a logical function on binary digits (BITs) of input information and provides a result in its output (O,2).

DEB Data Event Block. Memory control locations that have information accessed by programs, to provide control of asynchronous processing. The information is modified to indicate the progress of a particular event.

DEC DECimal. (a) A number-representation system with a number base of ten (i.e. the notation is represented by the characters 0 to 9). (b) An attribute or property which involves a condition, selection or choice from ten possibilities.

DECIT DECimal digIT. A particular character representing a whole number (not a fraction or mixed). In decimal (DEC) notation one of the characters 0 to 9.

DECR DECRement. The progressive lessening or decay of a free oscillation due to the expenditure of energy. It is the ratio of two successive peak values of oscillation of the same sign.

DEF Ministry of DEFence specifications. Until recently in the UK the main specification authority was the Ministry of Defence with their DEF specifications. However, these are gradually being superseded by the British Standard 9000 (BS 9000) series of standards for electronic components.

DEL (1) DELete. The action of removing or eliminating information, normally the removal of a record from a master file or the removal of a computer program from a memory after use.

(2) DELete character. A control character synonymous with back-arrow used to indicate that a previous character is to be ignored.

(3) DELay. (a) The time interval between the initiation of a signal and its reception; the time a pulse takes to traverse an electronic circuit or device. (b) In a computer the time taken to prepare or process data. (c) In data communications, the slowing down of information in a channel (CHNL(f)) for a specified period of time.

(4) Direct Exchange Line. Linked to only one subscriber, this single line has only one exchange number. Also known as 'exclusive exchange line' (EEL).

DELCL DELete CLause. A learn as you go aid in the programming in logic (PROLOG) computer programming language in the form of a built in command. It deletes clauses from the memory of the computer. Note add clause (ADDCL).

DEM DEModulator. Used in communications, a demodulator is a circuit or piece of apparatus that extracts an original signal (or intelligence) from a modulated carrier wave with the minimum of distortion. *Figure D.4* shows a simple demodulator circuit.

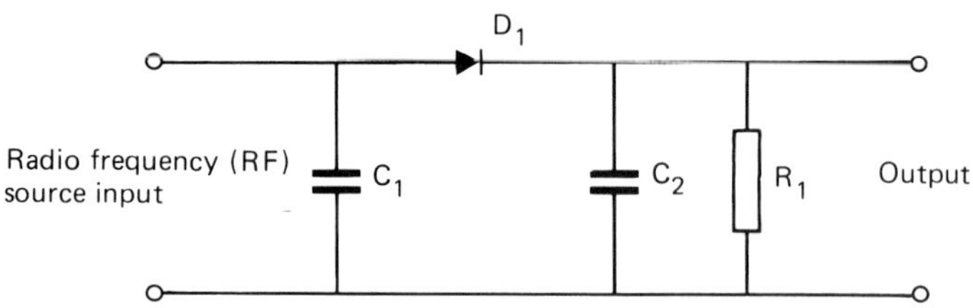

Figure D.4 A simple series-diode (D,5) demodulator (DEM) circuit

DEMOD DEMODulator. *See* DEM, above.

DEMUX (1) DEMUltipleX. The division of one or more information streams into a larger number of streams.

(2) DEMUltipleXor. (DEMUltipleXer, US.) A unit or piece of equipment that has a demultiplex (DEMUX,1) ability or facility.

DES Data Encryption Standard. Used for the protection of sensitive data this security algorithm specifies implementation in hardware for maximum security. Data is passed through a complex circuit by means of a user-supplied key, outputting a cipher text that is a non-linear function of the original data which is then virtually impossible to regenerate without knowing the key.

DESC Defense Electronics Supply Center. An American term used for a centre that controls the procurement policies and monitors the quality of military electronics contracts.

DETAB DEcision TABle. Used for the analysis of various problems, a DETAB contains a list of the possible courses of action, alternatives or selections that can be made.

DEU Data Exchange Unit. Used in communications, a DEU accepts data from a number of input channels (CHNL (f)), and sorts it according to priority and destination, at the same time performing checking and error routines.

DEUCE Digital Electronic Universal Calculating Engine. Built in the UK this early computer superseded the Automatic Computing Engine (ACE).

DEVCO DEVelopment COmmittee. A council committee of the International Standards Organisation (ISO) aimed at the needs of developing countries.

DF (1) Double Frequency. A two-way radio device employing a separate and distinctly different frequency in each direction.

(2) Direction Finder. A system employed to determine the direction of origin of a transmitted signal. It uses a radio receiver and a directional antenna in combination.

(3) Diversity Factor. The ratio of the sum of the maximum demands from a group of consumers to their maximum collective demands.

DFB Distributed FeedBack. A type of injection laser (LASER), whose wavelength of radiative longitudinal-mode oscillations is determined by the corrugation period of the structure and not by its material composition.

DFS Disk Filing System. A Collection of files contained on one or more magnetic disks, basically a computer filing cabinet.

DFT Diagnostic Function Test. A particular computer program employed to test the reliability of a system.

DGT (1) DiGiT. Any one of the symbols of integral value used in a system of numbering, each individual digit is used to designate a quantity smaller than the number base itself. In the decimal system the ten digits are 0, 1, 2, 3, 4, 5, 6, 7, 8, 9; in the binary system the digits are 0 and 1.

(2) Directorate General of Telecommunication. The postal, telegraph and telephone authority (PTT) of France.

DH Double-Heterostructure. A structure of injection laser (LASER), in which GaAsAl layers are fabricated on both sides of the gain region, which achieves reductions in the threshold current level. The device is capable of strong optical waveguiding and current confinement on both sides of its p-n junction (p-n).

DHE Data-Handling Equipment. A system of devices that can be either automatic or semi-automatic, used for the collection, transmission, reception and storage of information.

DI DeIonization. The process of removing dissolved salts or minerals from process water by ion exchange.

DIANE Direct Information Access Network for Europe. The information services offered over the European packet switching network (EURONET) system.

DIB Data Input Bus. Some microcomputers have a DIB, a mechanism in which data and address information is transferred between the switch register (SR) and the processor, also between processor and the memory, the memory and the input/output (I/O) interface and the processor and the input/output interface.

DID Direct Inward Dialling. A condition that allows a telephone caller direct access to any particular public automatic business exchange (PABX) extension without involving the exchange operator. Also known as direct dialling-in (DDI).

DIDS Domestic Information Display System. A computer graphics system that displays (D,6) the whole of the USA in the form of colour-coded maps.

DIGICOM DIGItal COMmunications system. A collection of units and devices which transmit information from one point to another in a digital (D,4) form.

DIL Dual-In-Line. A concept that provides for the standardized location and connection of components and integrated circuits (ICs). It is based on two rows of connections spaced at 0.1 inch pitch. The term can be applied to the pins on a dual-in-line package (DIP) or to the holes in a socket or printed circuit board (PCB).

DIM DIMension. A keyword used in beginners all-purpose symbolic instruction code (BASIC). An array is a list of items with a common name. To use an array it is necessary to say what size of array is required. This is done with keyword DIM.

DIMS Distributed-Intelligence Microcomputer System. A

method of multiprocessing which assigns to different processors fixed tasks allocated by software algorithms. The tasks include input/output (I/O) controller activities, data concentration, information processing, and remote communications.

DIN Deutsches Institut für Normung. (a) The West German specification system that can ostensibly be considered only as a guide to mating interface dimensions and tolerances. (b) Multipin audio and radio frequency (RF) connectors based on the West German standard are referred to as DIN connectors.

DIP Dual-In-line Package. A standard integrated circuit (IC) package. It has two rows of connection pins which extend down at 0·1 inch intervals. These packages are available in plastics for economy and ceramic for high humidities and temperature.

DIV (1) DIVider. A circuit that reduces the number of pulses or cycles by an integral factor. There are two basic types of dividers – analogue and digital (D,4).

(2) Integer DIVision. An operating word in the common algorithmic language (COMAL) computer programming language used for integer division.

DIVA Data Input-Voice Answerback. A system of communications where a user inputs, via a data terminal, to a computer and receives a voice answerback output (O,2) of a synthesized or recorded voice.

DJNR Dow Jones News/Retrieval. Up-dated daily this is a computerized information retrieval system covering business news and articles in the USA from leading financial periodicals.

DL (1) Data Link. Equipment that permits the transmission of information between two or more data stations.

(2) Delay Line. A transmission line, circuit, device or equivalent network that has been designed to introduce a known delay (DEL,3(c)) in the transmission of a signal.

DLC Data Link Control. A discipline or protocol comprising a set of rules used by computers or terminals when communicating, in order to ensure the orderly transfer of information.

DLD Dark-Line Defect. Luminescent areas that cross an active area during lasing and are responsible for gradual laser (LASER) degradation.

DLE Data Link Escape. The withdrawal of, or disestablishment of, a communication circuit that is used for the transmission of data in a coded form, in conjunction with data processing equipment (DPE).

DLEC Data Link Escape Character. The use of a control character combined with one or more succeeding characters to form an escape sequence and the development of extra data communications control operations.

DLL Dial Long Lines. In the USA, a range of equipment used to enable satisfactory signalling, supervision and/or transmission to be obtained on a subscriber's line that has a loop resistance higher than normal in the exchange equipment.

DLT Decision Logic Table. A concise representation of procedures or results of alternative courses of action as specified for varied groups or sets of combinations of conditions.

DM Delta Modulation. A form of modulation in which an analogue signal is sampled, and the samples are quantized and converted by coding into a digital (D,4) signal. The value of a sample of the signal is compared with an integrated value of the previous sample, and a positive binary pulse is transmitted if the analogue sample is greater than the reconstructed waveform or a negative pulse if it is smaller (*also see* DPCM).

DMA Direct Memory Access. A technique employed to provide the high-speed transmission of data between a computer peripheral and the computer memory without having to go through the central processor unit (CPU). Data is exchanged at maximum memory speed and several access methods to the memory are possible.

DMAC Direct Memory Access Controller. A circuit or device, now available on a single chip, providing control over applications that require data to be transferred into a computer memory rather than going through the central processor unit (CPU). The DMAC can improve performance in larger systems with a number of peripherals, by completely controlling block transfers between input/output (I/O) circuits and the system memory.

DMC Digital MicroCircuit. A miniaturized electronic circuit fabricated with inseparable interconnections and elements. It has been designed to accept at its input terminals particular logic states. These are converted according to set logic equations or function tables, and passed to the output (O,2) terminals.

DMCL Device Media Control Language. Software used in a computer for specifying the physical use of a data base logical data structure.

DME Digital Multiplex Equipment. This piece of equipment contains a digital multiplexor (MUX,2) and a digital demultiplexor (DEMUX,2). Also known as 'muldex' or 'muldem'.

DML Data Manipulation Language. Not a complete computer language in itself, DML employs a host language which it then uses as a framework. It is used by a programmer to communicate between an application program and a data base.

DMM Digital MultiMeter. A digital multimeter essentially converts an analogue, or continuous, signal into a series of digital (D,4) representations of sections of that signal. It ensures that the resultant display (D,6) is an accurate assessment of the signal amplitude. The DMM with its wider range of functions has largely superseded the DVM (digital voltmeter).

DMOS or **D-MOS** Double-diffused Metal-Oxide Semiconductor. A short-channel device that is produced by a double-diffused process in which the length of the channel is determined by the difference in lateral diffusion of two impurity distributions, as shown in cross-section in *Figure D.5*. DMOS devices exhibit excellent low-power and high-speed capabilities. *See* MOS.

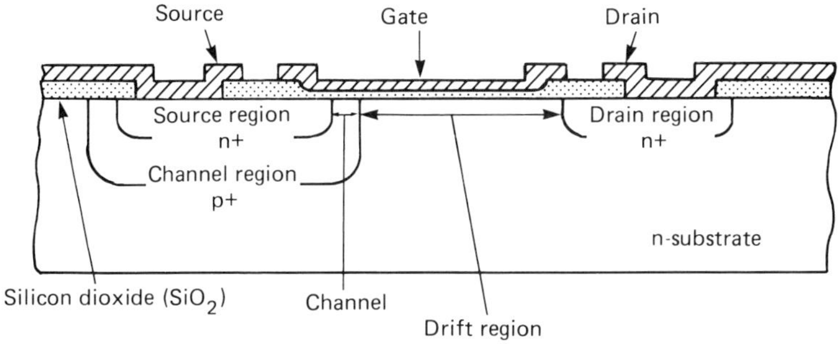

Figure D.5 Cross-section of a double-diffused metal-oxide semiconductor (DMOS) transistor with a channel formed by double diffusion in the source region

DMS Digital Multiplex Switching system. Providing a digital (D,4) circuit switched service for voice and data transmission, the DMS family of devices make use of pulse code modulation (PCM) and time division multiplexing (TDM).

DN Decimal Numbering system. A system based on 10 or the powers of 10 using the digits (DGT,1) 0–9 to represent numerical quantities.

DNA Digital Network Architecture. A particular design of construction for distributed computer systems developed by the Digital Equipment Corporation, who implement this architecture under the name DECNET.

DNC Direct Numerical Control. The system created by a number of numerical control (N/C) machines being connected to a common memory. The memory provides data, from stored machine programs, on demand to the NC machines.

DNL Dynamic Noise Limiter. A circuit which can be incorporated into the playback amplifier (AMP,1) of a tape recorder in order to reduce tape noise when it is most noticeable, particularly during quiet musical passages.

DO Digital Output. An output (O,2) signal, representing the magnitude of an input signal, that is produced in the form of a series of discrete quantities coded to represent digits (DGT,1).

DOD Direct Outward Dialling. The condition that allows CENTREX or PABX extensions to dial an external number without requiring the assistance of the particular exchange's operator.

DOS Disk-Operating System. An organized collection of techniques and procedures for operating a computer, as software (programs or routines), held on a disk.

DP (1) Data Processing. A general term describing the organization of data in a desired manner by automatic or semi-automatic means. Also sometimes a generic term for computing in general.

(2) Distribution Point. The final point at which the conductors forming an individual telephone subscriber's circuit in a local telephone line network are run to the subscriber's premises. Known in the USA as 'ready access terminal'.

(3) Dynamic Programming. A mathematical term that in essence requires that an optimum decision must be made at every stage of a multistage problem.

DPC (1) Data Processing Centre. A collection of automatic data-processing (DP, 1) equipment systems along with operators, in one centralized area under a single management group.

(2) Direct Program Control. An input/output (I/O) operation that allows direct control of a device by software programs.

DPCM Differential Pulse Code Modulation. A form of the process in which an analogue signal is sampled, and the samples quantized and converted by coding into a digital (D,4) signal. The actual value of every sample and its predicted value are quantized and converted by encoding into a digital signal.

DPDT Double-Pole Double-Throw. A particular construction of switch which can be incorporated into a double line or pole. It has two centre or common terminals each having a moveable blade connected so that each simultaneously makes contact, when activated (thrown) in one direction or the other, with its respective terminal of one of the two pairs of terminals available. *See Figure D.6.*

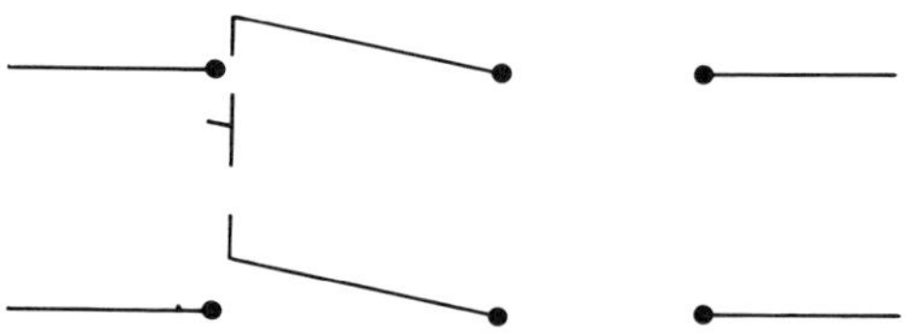

Figure D.6 Symbol for a double-pole double-throw switch (DPDT)

DPE Data Processing Equipment. Systems comprised of interconnected components and sub-assemblies used for data processing (DP,1).

DPI Device Programmer Interface. An interface between a computer and a programmer, downloading compiled program tables in either Joint Electron Device Engineering Council (JEDEC) or high level (HL) format, or uploading the tables of existing devices for editing or simulation.

DPM (1) Data Processing Machine. Generally refers to a machine that can store and process numeric and alphabetic information.
(2) Digital Panel Meter. A meter that provides in digital (D,4) form a display (D,6) of any particular quantity or event that has been measured or recorded. Two preferred types are light-emitting diode (LED) displays for their self-illumination and contrast, and liquid crystal displays (LCD) for their low power consumption.
(3) Documents Per Minute. A measure of the rate at which media with information recorded, normally paper with printing, can be processed by equipment such as data processing equipment (DPE).

DPS (1) Data Processing System. A number of components and sub-assemblies connected together to receive, transmit, or store data. Many systems can also process information to a set protocol of mathematical formulae and produce desired results.
(2) Data Processing Standards. Written guide lines and procedures laid down for data processing (DP) as well as for other aspects of computing. A number of British Standards Institution (BSI) committees were concerned with these standards, which have now been brought into line with corresponding International Standards Organisation (ISO) standards.

DPST Double-Pole Single-Throw. A particular construction of switch which can be incorporated into a double line or pole. It has two terminals each having a moveable blade. The blades are connected so that each simultaneously makes contact, when activated (thrown), with its respective line terminal. *See Figure D.7.*

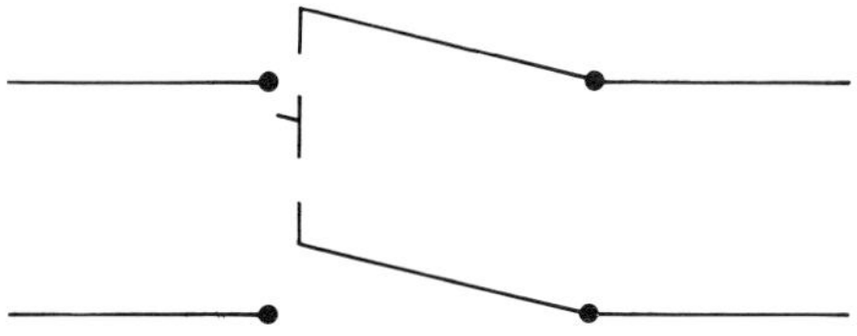

Figure D.7 Symbol for a double-pole single-throw switch (DPST)

DRAM Dynamic Random-Access Memory. Called 'dynamic' because information is stored only temporarily, the dynamic random-access read/write memory (RAM) must be continually rewritten or refreshed. Storage is made in the form of the parasitic leakages that are experienced in integrated circuits (ICs). Although the transitory nature of its storage is a drawback, the saving grace of a DRAM is its small size. *See Figure D.8.*

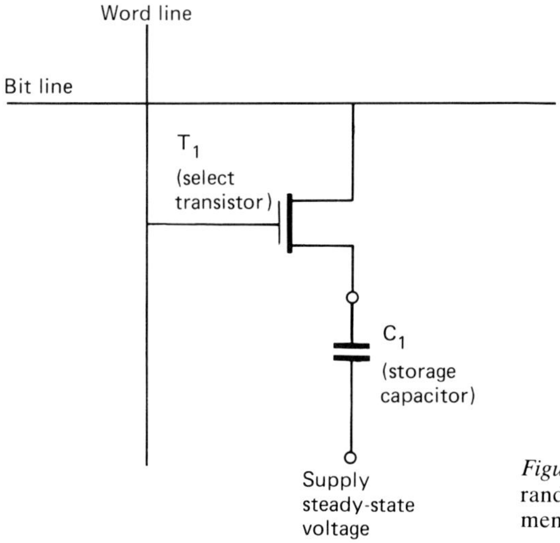

Figure D.8 A basic dynamic random-access read-write memory (DRAM) cell

DRAW Direct Read After Write. In an optical digital (D,4) disk recording system, information once written cannot be removed. DRAW permits the recognition of errors enabling them to be corrected by rewriting data in a new position on the disk, erasing the location of the incorrect information from the memory of the computer.

DRC Design Rules Checking. The checking of design data for user-selected types of errors and manufacturing tolerance violations. Uscd in computcr-aided design/computer-aided manufacture (CAD/CAM).

DRL Data-Retrieval Language. A computer language used in an information retrieval system, whereby information can be searched for particular items, in response to the definition of particular user needs.

DRO Destructive Read-Out. The phenomenon of data being lost when it is read from a storage device, all recording of it being destroyed.

DRS Document-Retrieval System. An information-retrieval system in which the contents of a document are indexed by means of key-word-in-context (KWIC) terms.

DS Data Set. A compilation of similar and related data records that have been recorded for use by a computer.

DSA Data Set Adaptor. Used to interface a computer to a modulator demodulator (MODEM). It reduces groups of binary digits (BITs) into individual bits for serial transmission and reverses the process for the reception of signals.

DSB Double SideBand transmission. A form of amplitude modulated (AM) transmission used in sound broadcasting in which frequency bands are produced above and below the carrier wave. The upper sideband is the frequency band occupied by the side frequencies above the carrier frequency. Similarly the lower sideband is the frequency band occupied by the side frequencies below the carrier frequency.

DSC District Switching Centre. In a four-wire-switched network, that enables any two group switching centres to be interconnected by not more than five transit links, this is the designation given to a switching centre connected to a number of group switching centres (GSC) and at least one main switching centre (MSC).

DSCB Data Set Control Block. A storage location, containing information necessary for the control of one data set (DS), that is needed to describe and manipulate the data set in a direct-access (DA,2) device.

DSE Data-Switching Exchange. A particular section of a computer network, which allows the transmission of discrete packages of information. It performs switching and routing functions and normally consists of a mini- or a microcomputer.

DSECT Dummy SECTion. Part of a computer program that translates a source program written in assembly language into object code. It contains data declarations but does not reserve space for that data.

DSI Digital Speech Interpolation. A technique to increase the capacity of long-distance digital (D,4) systems used for the transmission of speech. When a person is silent no binary digits (BITs) are transmitted; when the speech begins again, bits flow. It is used in INTELSAT satellite communications.

DSL Data Set Label. The identification given to a storage location containing information necessary for the control of one data set (DS).

DSR Data-Set Ready. A line on a modulator demodulator (MODEM) which informs the transmitting terminal that the signal received is all right. An RS-232 standard (*see* RS-232C).

DSV Digital Sum Variation. An expression of the difference between the maximum and minimum values of the digital (D,4) sum of a coded signal.

DSW Device Status Word. A computer word that contains binary digits (BITs), the condition of which indicates the particular status of various devices.

DT Data Transmission. (a) The sending of data by means of signals sent over a telecommunication network, from a data source to a data sink. (b) The sending of data from one part of a system to another.

DTE Data Terminal Equipment. (a) A unit at a data station that can receive transmitted data signals and/or transmit data. It can also control information transfer according to the protocol of the data link network. (b) Any piece of equipment at the beginning or end of a communications path. It can be a terminal or computer or the end user of a telephone circuit.

DTF Dynamic Track Follower. A system for video recording and playback which ensures that automatic realignment of video heads relative to their track during playback.

DTL Diode-Transistor Logic. A combination of diodes and transistors in an integrated circuit (IC) form that comprises the logic elements of a system. A basic DTL logic gate design is shown in *Figure D.9*.

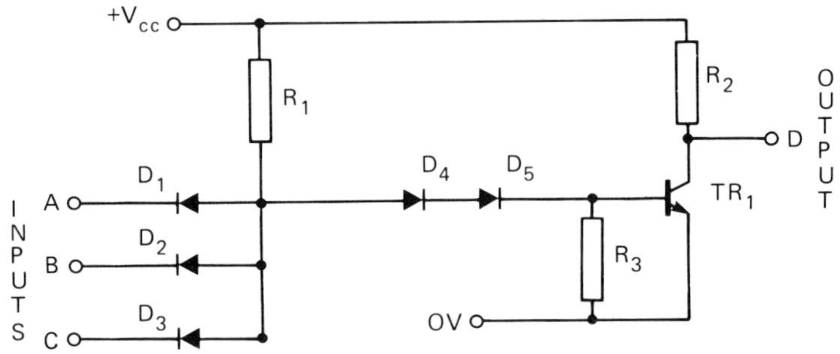

Figure D.9 A basic diode-transistor logic (DTL) gate design

DTMF Dual-Tone MultiFrequency. A name given to telephone pulsing where every digit is represented by a particular pair of audio frequencies, one below 1000 Hz, the other above 1200 Hz.

DTR Data Terminal Ready. A line on a terminal which informs the modulator/demodulator (MODEM) that it is ready to transmit data.

DTS Data-Transmission System. A series of devices, circuits or modulator/demodulators (MODEMs), that can translate or transfer information from one position to another.

DTSS Dartmouth Time Sharing System. An interactive computer system developed at Dartmouth College in the USA which was operated in 1969. The system enables non-computer science specialists to use the system, as it provides user-friendly facilities.

D-type (1) A connector employed in both the data processing (DP,1) and telecommunication (TELECOMM) industries. Available in sizes from 9 to 37 ways, it is usually used for interconnecting between equipment or for connecting different assemblies within one unit. The name derives from the elongated D shape of the pin housing.

(2) A flip-flop (or bistable) switching circuit that has two inputs D and CK (for clock) and two outputs Q and $\bar{Q}$, *see Figure D.10*. The output Q takes up the state of the input D when a pulse is applied to the clock input. Output $\bar{Q}$ goes to the opposite state.

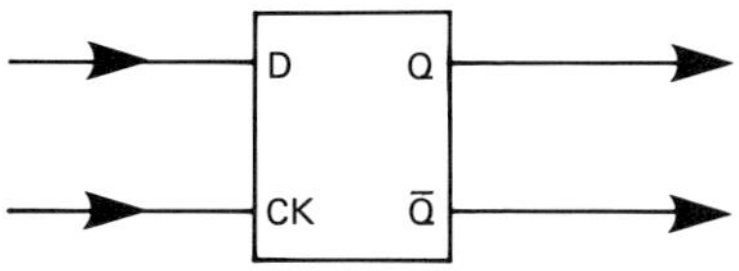

Figure D.10 The D-type flip-flop symbol

DUP DUPlicate. A stack operator employed in the FORTH computer programming language. Its sole function is to recognize the elements on the stack. It duplicates the top item on the stack, for example:

5 DUP yields 55

DURAL DURALumin. An alloy of aluminium containing copper, manganese and silicon and having a high tensile strength. It has many uses where conduction and movement are required, e.g. switchgear moving parts.

DUV Data Under Voice. A system of digital (D,4) data transmission carrying information on a frequency below that used for voice transmission. *See* DAV (2).

DVM Digital VoltMeter. A meter that displays (D,6) in a digital (D,4) form, measurements of alternating current (AC) or direct current (DC,1) voltages. It has generally been superseded by the digital multimeter (DMM).

DWBA Direct-Wire Burglar Alarm. A wired electrical circuit used for the transmission of alarm signals from a protected premises to a monitoring station, often a police station.

DWM Destination Warning Marker. A warning spot placed on magnetic tape (MAGTAPE) about 2 metres from its end. The spot is reflective and can be sensed photoelectrically indicating the proximity of the end of the tape.

DX (1) DupleX. A term describing the simultaneous transmission over a channel (CHNL(a)), link or circuit in both directions.
(2) DupleX signalling. A signalling system which occupies the same cable pair as the voice path, but does not require any special type of filtering.
(3) Long distance communication. The success achieved when DXing (long-distance radio listening and often two-way communications) depends substantially on prevailing propagation conditions, which include local weather conditions, various atmospheric conditions and sunspots.
(4) Distance. The number of corresponding digit positions which differ between two binary words of the same length. The distance of the two binary words 11001101 and 10110011 is 6.

DYSTAL DYnamic STorage ALlocation. A formula translator (FORTRAN)-based system computer language. When used, the elements of a list are kept in consecutive storage locations so they can be integer-indexed.

E

e Emitter of a transistor. The region of a bipolar transistor from which charge carriers flow into the base (b) region.

E (1) Exa. A decimal prefix used in association with a base unit in the Système International d'Unités (SI) system of units indicating a multiplication factor of 10^{18} of that unit. For example, one exatonne, Et, is equal to 1 000 000 000 000 000 000 tonnes.
(2) Erlang. A unit expressing intensity of traffic. One continuously-engaged line has a traffic flow of one erlang; if the average number of calls made at any one time is 12, then the traffic intensity is 12 erlangs.
(3) The hexadecimal (HEX) symbol for binary 1110, decimal 14.
$E_{16} = 14_{10} = 16_8 = 1110_2$.

EA Effective Address. The location or unit in a computer, where information is stored that is actually used for the execution of an instruction. This may differ from that of the instruction in storage.

EAE Extended Arithmetic Element. A central processor element in a microprocessor that is implemented by hardware to multiply, divide and normalize functions.

EAM Electrical Accounting Machine. A term that can be applied to any one of the many electro-mechanical devices employed in data processing (DP,1). It can include tabulators and calculators but excludes digital (D,4) electronic devices.

EAN European Article Number. A computer coded number in the form of a bar code. Bar codes are used on consumer products as identification.

EAPD ElectroAbsorption PhotoDiode. An avalanche photodiode made from a direct-gap semiconductor material, like gallium arsenide (GaAs), having a low carrier concentration. The responsivity and quantum efficiency are high at wavelengths close to and beyond the absorption edge in GaAs. The avalanche gain at these wavelengths is much higher than the gain at shorter wavelengths.

EAPROM Electrically Alterable Programmable Read-Only Memory. A specialized read only memory (ROM). The contents can be erased *en masse* in one operation by applying a higher than normal voltage to a program input. The information held in the memory can be altered completely or partially, some special types allow a single word to be erased and rewritten. These nonvolatile memories have no fusible links (FL) and do not require ultra-violet (UV) irradiation.

EAROM Electrically-Alterable Read-Only Memory. A specialized random-access read/write memory that can be programmed by writing into the array and then generally used as a read-only memory (ROM). The devices have a fast read cycle of around 15 microseconds (μs) and a much slower write cycle of about one millisecond (ms). Other qualities are as EAPROMs (*see above*).

EAX Electronic Automatic eXchange. A telephone exchange employing techniques such as digital (D,4) switching, in which communication between subscribers is effected without the intervention of exchange operators.

E/B Electrode per Bit. A form of operation of a charge-coupled device (CCD) memory that increases the number of binary digits (BITs) of information that can be stored.

EBCDIC Expanded (or Extended) Binary Coded Decimal Interchange Code. An 8-binary digit (BIT) code, similar to American Standard Code for Information Interchange (ASCII), that uses the eighth level as an information bit. The code is used to represent 256 unique letters, numbers and special characters.

EBIC Electronic Beam Induced Current. A method of contactless testing employed in automatic test equipment (ATE), in which a localized current is induced in a semiconductor device in order to examine the internal logic.

EBL Electron Beam Lithography. A semiconductor fabrication process very similar to photoengraving. It is a method for producing patterns of etching, doping, metallization or insulation on the surface of a semiconductor material, slice or wafer. The process is normally applied to the chromium-on-glass mask used in the production of integrated circuits (IC), but can be used directly on the slice or wafer.

EBR Electron Beam Recording. A process that employs an electron beam to store and read information on a particular target. These targets are normally silicon dioxide and electrostatic charges actually store the data. On material such as vitrium iron garnet the data is stored as magnetic bubbles.

EBU European Broadcasting Union. An organization of radio and television (TV(a)) broadcasting authorities from a number of European countries.

EC Error Correcting. The process of detecting invalid characters that will not conform to a set of rules. The construction of the rules is held in an error-correcting code.

ECAP Electronic Circuit Analysis Program. A computer language used for the modelling and analysis of electrical networks. It employs function generators or tables of functions along with passive components for close examination of device models.

ECB Event Control block. A group of words or records treated as an individual unit that indicate the status of predictable events and take appropriate action where necessary.

ECC Electronically-Controllable Coupler. In fibre optics (FO), this is part of an optical system which enables other parts to be coupled or uncoupled from each other via an electrical signal. (2) Earth Continuity Conductor. Any conductor which can provide the continuation of an earthing or ground circuit, e.g. the earth in an electric cable, which can be sheathed or bare, can be referred to as an ECC.

ECCM Electronic Counter-CounterMeasures. Electronic equipment, devices and techniques employed to destroy or degrade the effectiveness of an enemy's electronic countermeasures (ECM).

ECD ElectroChromeric Display. Low-voltage, low-current devices that are chemically transformed from transparent to opaque when excited by an external electric field. When the field is turned off the display (D,6) continues to be opaque until subjected to a field with a polarity opposite to the first. The displays created are legible over a wide viewing angle and are compatible with integrated circuit (IC) technology.

ECG ElectroCardioGraph. An instrument which produces a trace (called an 'electrocardiogram') that provides a record of the voltage (V,2) and current (I) emanating from the hearts of living animals or humans.

ECH Eddy-Current Heating. Heat produced in a conductive material, induced by the material being subjected to a varying magnetic field (eddy-current) usually from a coil with an alternating current (AC,1).

ECHO European Commission Host Organization. An organization that makes available a computerized referral and enquiry service, as well as providing information to a number of stores of computer files.

ECL Emitter-Coupled Logic. A family of circuits that have been designed to perform a specific logical function. Pairs of transistors that form the logic circuit are connected together by their emitters, as shown in *Figure E.1*.

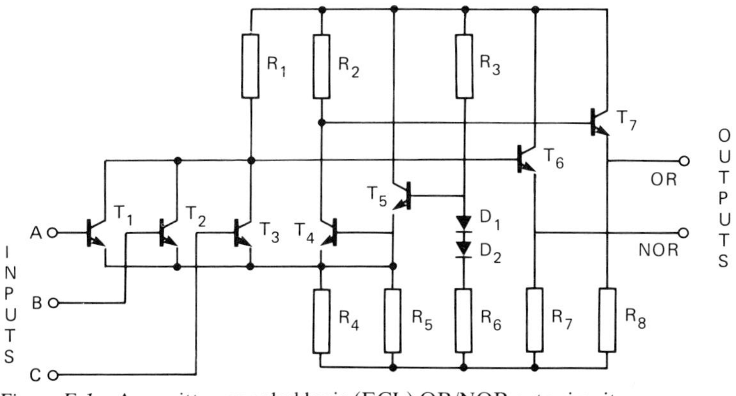

Figure E.1 An emitter-coupled logic (ECL) OR/NOR gate circuit

ECM Electronic CounterMeasures. Electronic jamming equipment and other devices and techniques used in tactical and strategic applications to destroy the effectiveness of an enemy's electronic aids to warfare, as a function of electronic warfare (EW).

ECMA European Computer Manufacturing Association. An organization comprising original equipment manufacturers (OEM) of computers from a number of European countries.

E-COM EleCtrOnic Mail. A service for the electronic transmission or distribution of messages, provided by the US postal service.

ECS (1) European Communications Satellite. Intending to provide Europe with a satellite communications system, this

programme is run by the European Space Agency (ESA).
(2) Experimental Communication Satellite. A range of experimental communication satellites put up by Japan.

EDA Electronic Differential Analyser. An analogue computer, using a number of interconnected integrators, employed in solving differential equations.

EDP Electronic Data Processing. The conversion of data by electronic equipment into a form able to be processed by computers. EDP is a term used more often in the USA whilst data processing (DP,1) is more often used in the UK and Europe.

EDPM Electronic Data-Processing Machine. An individual unit or device that either is part of an electronic data processing system (EDPS), or can provide stand-alone electronic data-processing (EDP).

EDPS Electronic Data-Processing System. A collection of electronic data-processing machines (EPDMs) that are capable of performing electronic data-processing (EDP) functions.

EDS Exchangeable Disk Store. A generic term used to describe storage disks that are capable of being removed from a computer and replaced by others.

EDSAC Electronic Delay Storage Automatic Computer. Developed in 1949 by the mathematics department of the University of Cambridge, EDSAC was one of the first electronic computers.

EDVAC Electronic Discrete Variable Automatic Computer. One of the first electronic computers, designed and built in 1949 at the University of Pennsylvania.

EEC European Economic Community. Commonly known as the 'Common Market', the EEC was founded by the treaty of Rome in 1957, and was extended in 1973, 1983 and 1986. Its membership includes Belgium, Denmark, France, West Germany, Greece, Ireland, Italy, Luxembourg, Netherlands, Portugal, Spain and the United Kingdom. The headquarters are in Brussels.

EEG ElectroEncephaloGraph. An instrument which produces a trace (called an 'electroencephalogram') that provides a record of the voltage (V,2) waveforms which emanate from the brains of animals and humans.

EEL Exclusive Exchange Line. Also known as a direct exchange line, this is a single telephone line that provides service to an individual subscriber and has only one exchange number.

EEM Electronic Engine Management. A microprocessor-based system that provides monitoring for cost-effective running of a particular vehicle. An EEM system would automatically control various fuel input parameters and control engine speed, resulting in improved fuel consumption and engine efficiency.

EEPROM Electrically-Erasable Programmable Read-Only Memory. Using only silicon and its derivatives plus metals, this device employs the floating-gate tunnel oxide (FLOTOX) cell structure in which electrons tunnel to and from a floating gate. *See Figure E.2.* It is a field programmable read-only memory

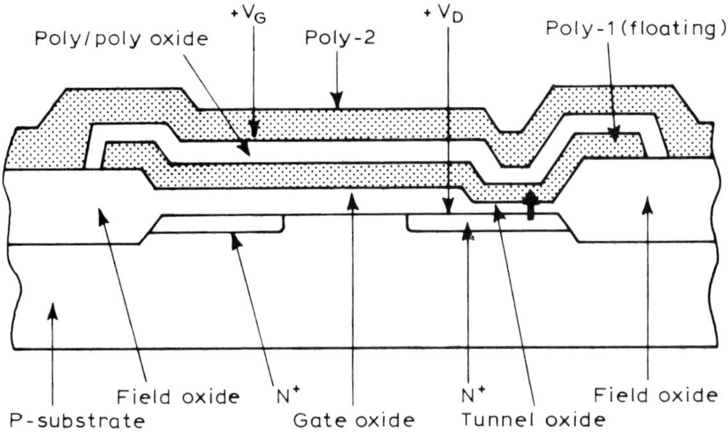

Figure E.2 A cross-section of an electrically-erasable programmable read-only memory (EEPROM) that employs the floating-gate tunnel

(FROM) in which the cells may be erased electrically in one second and may be reprogrammed electrically. It can also be erased and reprogrammed up to a million times. This memory device in general makes practical such applications as altering remotely, by telephone, microprocessor system programs. Also known as E^2PROM.

EET Equipment Engaged Tone. A tone heard by a caller if any part of a telephone network is overloaded and the call cannot be completed. An engaged tone (ET) or busy tone (BT,2) is used for this purpose in many systems replacing an EET, although the busy tone actually indicates that a called party is busy.

EFL Emitter Follower Logic. A type of bipolar integrated circuit (IC) fabrication technology used in logic circuits.

EFT Electronic Funds Transfer. The withdrawal of funds from one account to be deposited in another by electronic means. *See* TELEBANKING.

EFTA European Free Trade Association. A collection of trading countries within Europe set up in 1960 to promote economic development and expansion of trade. Members are Austria,

Finland, Norway, Portugal, Sweden, Switzerland and Iceland. The headquarters are in Geneva.

EFTS Electronic Funds Transfer System. A general term used in the USA that refers to various electronic communications systems that are able to transfer financial information from one point to another. The term can encompass automated clearing houses and point-of-sale (POS) systems.

EHF Extremely-High Frequency. Frequencies occurring between 30 GHz and 300 GHz, forming part of a larger continuous range of frequencies. EHFs are considered by international agreement to be radio frequencies in band number 11 and have wavelengths of between 1 mm and 1 cm.

EHOG European Host Operators Group. An organization representing the interests of a number of information vendors, who hold various stores of computer files, within Europe.

EHT Extra-High Tension. (a) Generally a term used to describe the high-voltage supply for display (D,6) tubes. The displays can be either cathode-ray tubes (CRTs) or television (TV(c)) picture tubes. (b) The term is also part of power distribution terminology.

EHTPS Extra-High-Tension Power Supply. A form of power-supply unit (PSU) that can provide a high voltage (HV,2), normally several thousand volts (kV) output (O,2), from a lower voltage input. *See* EHT.

EIA Electronics Industries Association. A trade association of the electronics industry in the USA. It formulates technical standards, such as the widely used RS-232C interface standard for connecting computer equipment, and also maintains contact with American government agencies in matters relating to the electronics industry. It is one of the bodies that help determine standards for the USA Standards Institute (USASI).

EIAJ Electrical Industries Association of Japan. A trade association of the electrical industry in Japan.

EIES Electronic Information Exchange System. Set up in the USA, this experimental computer network is supported by the National Science Foundation. Connecting a number of groups together, it provides electronic mail (E-COM), teleconferencing and personal electronic communications, whilst monitoring their use and impact.

EIN European Informatics Network. A heterogeneous packet-switched network set up by agreement between a number of European countries.

EIRP Effective Isotropically Radiated Power. A comparison made to a given reference antenna, by the relative gain in a given direction for a particular power supplied to an antenna.

Known in the USA as 'effective radiated power' (ERP).

ELCB Earth-Leakage Circuit-Breaker or Earth-Leakage Contact-Breaker. A device that provides a disconnection or break in electrical supply if the voltage (V,2) in the earth or non-current-carrying part of a circuit, or the out-of-balance current in the supply due to leakage, exceeds a predetermined level.

ELD Electro Luminescent Display. An illuminating device used to backlight a liquid crystal display (LCD), so that it can be used in a darkened environment. The unit is contained in a plastics sleeve for protection from moisture, is available in a number of low-reflection colours providing uniform brightness and has a long dependable life.

ELED Edge-emitting Light Emitting Diode. A hermetically sealed device developed for use in high speed fibre optic (FO) digital (D,4) communication applications.

ELF Extremely-Low Frequency. A frequency less than 100 hertz (Hz).

ELINT ELectronic INTelligence. Electronic techniques employed in signal analysis and location as tools for electronic warfare (EW). These are combined with the interpretation of the information gained concerning an enemy's activities.

ELM Extra Low Mass. A name given to a range of record player arms that have good horizontal and vertical track-following capability.

ELSI Extra-Large-Scale Integration. The design and production of integrated circuit (IC) or microcircuit chips (CHIP) in which the solid state circuitry is manufactured on a single piece of material, normally silicon. ELSI refers to circuits with a capability of more than one megabit. Sometimes also referred to colloquially as 'Even-Larger-Scale Integration'.

ELV Extra-Low Voltage. In European regulations a term used to imply a voltage not exceeding 50 volts (V,1).

EM End of Medium. A specific control character used to indicate the physical end of a data medium. It normally allows sufficient time for the user to adjust or change the system of operation.

EMC ElectroMagnetic Compatibility. The ability of generators of electromagnetic radiation and equipment susceptible to interference from such radiation, to exist in the same environment without degradation of performance. This is achieved generally by reducing the level of radiated electromagnetic interference (EMI), by introducing attenuation in all EMI paths between generator and susceptible equipment.

EMF ElectroMotive Force. The physical property of a device that enables it to drive an electric current (I) around a circuit. The property can be produced by either mechanical or chemical means, for example a generator or a battery (BAT).

EMI ElectroMagnetic Interference. The distortion or impairment of a desired transmission or signal brought about by an electromagnetic disturbance.

EMI/RFI ElectroMagnetic Interference and Radio Frequency Interference. A range of conducted or radiated electromagnetic emissions that produce an unintentional and unwanted response in a piece of electronic equipment or a computing device.

EMIS Electronic Materials Information Services. A store of information held on computer files concerning material properties and material supply. It is produced by the Information Services: Physics Electrical and Electronics, and Computers and Control (INSPEC) body.

EMMS Electronic Mail and Message System. The services in a communications system or network such as electronic mail, teleconferencing, word processor communication and videotex.

EMP ElectroMagnetic Pulse. High-energy radiation, such as that produced by a nuclear blast.

EMS Electronic Message System. First used as a description of communication made using terminals in a communications network or system. It is now generally employed to describe specialized services such as electronic mail, teleconferencing, word processor communication and videotex.

EMW ElectroMagnetic Wave. The resultant effect obtained from the interaction of an electric field and magnetic field. The interaction causes energy to be transmitted in a particular direction.

EN ENamelled. A covering (insulation) often given to fine wires.

END A computer statement used to denote the end of a computer program.

ENG Electronic News Gathering. A range of portable television (TV(a)) broadcasting equipment.

ENIAC Electronic Numerical Integrator And Calculator. Considered by many to be the first truly electronic computer, ENIAC was produced by the University of Pennsylvania during the Second World War. It contained more than 18 000 electron tubes.

ENQ ENQuiry. A special control character used in data communications. It has been designed to elicit a response from a remote station or terminal. The response usually required is some form of identification and/or operating status.

ENR Excess Noise Ratio. A ratio of the noise output of a particular device against a noise temperature of 290° kelvin (K,2) (which is a defacto standard reference point).
ENR (in dB) = 10 × 10g10 (TE/TO − 1)
TE = Noise temperature of device
TO = 290°K.

EO End Office. Class 5 of five class numbers assigned to offices according to their function, used in the American direct distance dialling (DDD) telephone network. Any one centre can handle telephone call traffic from one, two or more centres lower in the hierarchy, Class 1 being the top classification.

E-O ElectroOptic. Relates to the conversion of electrical power to optical power (electrical signal to optical signal conversion).

EOA End Of Address. A specific control character used to indicate the finish of characters that are not text, i.e. they are address information, employed on a transmission line.

EOB End Of Block. A control character used to indicate the finish of a group of words or records treated as an individual unit, employed on a transmission line.

EOD End of Data. A computer program function used for detecting the end of data.

EOF End Of File. (a) A mark used at the end of a quantity of data to indicate its completion. (b) A computer statement that indicates the end of a file of information.

EOJ End Of Job. (a) A status condition that exists in a system when a particular job has been completed. (b) A computer statement that indicates the end of a task or job.

EOL End Of Line. A specific character that indicates the termination or finish of a group of records.

EOM End Of Message. A specific character or set of characters that indicate the termination or finish of a message or record.

EOR (1) End Of Reel. A character or word employed to indicate the termination or finish, in a multi-reel magnetic tape file, of all but the last reel.
(2) End Of Record. A computer statement used to indicate the finish of a record or a set of records.
(3) End Of Run. A computer statement used to indicate the finish of a computer run.

EOT (1) End Of Transmission. An indicator consisting of a unique group of characters at the termination of a transmission. It denotes the end of a data transmission to or from a remote terminal and also resets stations in the transmission line to control mode.
(2) End Of Tape. A special mark that can be either a specific code on the magnetic tape (MAGTAPE) or a physical indicator

used to show the physical end of a reel of tape. The physical indicators can be in the form of an easy-to-see reflective strip or a transparent section.

EOV End Of Volume. (a) A marker that indicates the finish of a physical volume. (b) A computer software routine that comes into operation and takes control when the situation in EOV (a) is indicated.

EPB Environmental Periodicals Bibliography. A store of information held on computer files, in the USA, that relates to aspects of environmental sciences.

EPIC Exchange Price IndiCation. A store of information held on computer files, managed by the London Stock Exchange. It contains information from a number of international sources.

EPLD Electrically Programmable Logic Device. A particular type of programmable logic device (PLD) that is capable of being erased electrically and reprogrammed electrically, such as electrically erasable programmable read-only memory (E^2PROM).

EPR Ethylene Propylene Rubber. A material used for cable insulation which has excellent electrical properties, good corona resistance and good low-temperature flexibility.

EPROM Erasable Programmable Read-Only Memory. A read-only memory (ROM) that can have information that is stored in its memory changed over and over again. *See Figure E.3.* EPROMs that use metal-oxide semiconductor (MOS) technology can be erased by ultra-violet (UV) light through a small window in the top of their packaging.

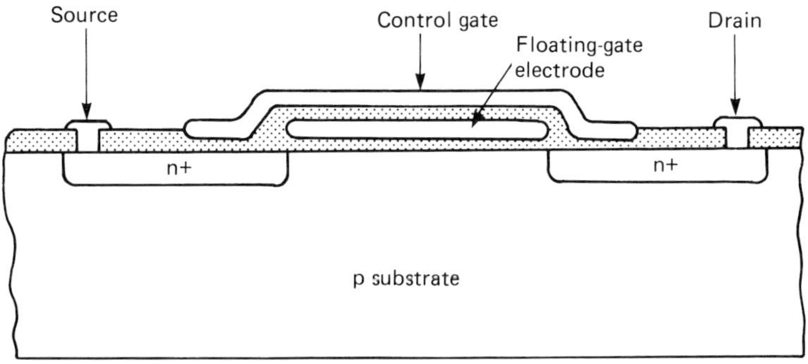

Figure E.3 A cross-sectional view of an erasable programmable read-only memory (EPROM)

E^2PROM Electrically-Erasable Programmable Read-Only Memory. Using only silicon and its derivatives plus metals, this device employs the floating-gate tunnel oxide (FLOTOX) cell structure in which electrons tunnel to and from a floating gate. *See Figure E.2.* It is a field programmable read-only memory (FROM) in which the cells may be erased electrically and may be re-programmed electrically. Also known as EEPROM.

EPSS Experimental Packet-Switched System. An experimental packet-switched computer network set up by the Post Office in the UK in 1979.

EQ (1) EQual to. A comparison of two values that states whether a second value is equal to the first. The symbol is =.
(2) EQualization. A deliberate reshaping of frequency response, employed in tape recorders in order to overcome problems of tape saturation.

ER Error Recovery. A system or procedure that has been designed to overcome the deviation of a computed or a measured quantity from the theoretically correct or true value.

ERP Effective Radiated Power (US). A comparison made to a given reference antenna, by the relative gain in a given direction for a particular power supplied to an antenna. Known in the UK as 'effective isotropically radiated power' (EIRP).

ESA European Space Agency. Responsible for the cooperative development of launch vehicles and satellites for scientific, weather and communication use, ESA was founded in 1975.

ESA-IRS European Space Agency-Information Retrieval System. With its headquarters in Italy, the European Space Agency (ESA) provides access to a number of stores of information held on computer files, relating to scientific matters.

ESC (1) EScape Character. A special symbol, part of a character set, employed to show that following characters belong to a different character set from those that have gone before.
(2) Edison Screw Cap. A lamp cap or bulb cap, with the outer case in the form of a screw thread which provides one contact, the other contact being a central end projection.
(3) Normally an integrated circuit (IC), the ESC holds the speed of a cassette drive motor constant, in the face of variations in voltage (V,2), temperature and cassette characteristics.

ESD ElectroStatic Discharge. A potentially damaging flow of static electricity. Such flows are especially dangerous to small junction area semiconductor devices (*see* static sensitive device (SSD)). To guard against, SSDs are transported with their pins or leads in conducting foam, and worked within special handling areas (*see* SHA and *Figure S.4*).

ESI (1) Externally Specified Index. Enabling a computer system to become a central message switching centre, for a variety of remote devices, this facility automatically routes messages to and from a main store. It achieves this without the program sequence of the central processor unit (CPU) being disturbed. (2) Electricity Supply Industry. Those who undertake the planning of cheap and reliable supply of electric power to those who desire it.

ESR Equivalent Series Resistance. The amount of resistance in series with an ideal or loss-less capacitor (C,6) which exactly duplicates the performance of a real capacitor. In general, the lower the ESR, the better the quality of the capacitor and the more effective it is as a filtering device. It is a prime determinant of ripple in switching power supplies.

ESS Electronic Switching System. Originally a term for a Bell System computerized telephone exchange, ESS is now used generally as an industry term.

ET Engaged Tone. In a telephone system that employs an equipment engaged tone (EET), an ET or busy tone (BT,2) indicates to a caller that the called party is engaged. In systems not employing EET, an ET (as in the UK) can indicate that any part of a telephone network is busy.

ETB End of Transmission Block. A control character employed to indicate the termination or finish of transmission of a group of words or records treated as an individual unit.

ETIM Elapsed TIMe. The period of time measured from the start to the completion of a process. This apparent time period can sometimes be longer than the time actually taken in processing information.

E-to-E Electronic-to-Electronic. An expression applied to video tape recording (VTR) where a signal is processed by electronic means, but not recorded.

ETV Educational TeleVision. Available to television (TV(c)) receivers in the home, this form of broadcasting informs and teaches. UK examples are schools' programmes and the Open University.

ETX End of TeXt. A specific control character employed to indicate the end of transmission of a piece of text.

EURODICAUTOM EUROpean AUTOMatic DICtionary. A machine aided translation (MAT) system employing a bank of specific terminology that can assist in the translation of scientific and technical documents.

eV Electron-Volt. A unit of energy that can be defined as the amount of energy released or acquired when an electric charge or one electron moves through a potential difference (PD) of one volt (V,1).

EVA Ethylene Vinyl Acetate. A material used to improve the handling capabilities of optical fibres (OF) protected by a thin (5 to 10 μm) plastics coating. The improvement is affected by the addition of a second, much thicker, coating (250 μm) of EVA.

EW Electronic Warfare. The use of electronic devices for the purposes of: determining the existence and disposition of an enemy's electronic aids to warfare; destroying or degrading the effectiveness of an enemy's electronic aids to warfare; preventing the destruction of the effectiveness of friendly electronic aids.

EVCB EVent Control Block. An area of storage used by a computer operating system (OS) to record the current condition or status or an event.

EVR Electronic Video Recording. The recording of video signals electronically on magnetic tape (MAGTAPE), magnetic disk or any other suitable material.

EXEC (1) EXECutive statement. A job control character that labels the load module to be executed along with specific job steps.
(2) EXECute. To carry out the operations specified by a particular instruction or routine.

EXTRN EXTeRNal reference. An indication of where to find a single variable from a range, or where to find an item which is not defined in a particular computer program, segment or subroutine.

F

f (1) Femto. A decimal prefix commonly used in association with a base unit in the Système International d'Unités (SI) system of units. It indicates a multiplication factor of 10^{-15} of that unit. For example, one femtosecond, fs, is equal to 0.000 000 000 000 001 second.
(2) Frequency. In the Système International d'Unités (SI) system of units, the unit of frequency (of a periodic phenomenon) is the hertz (Hz). It is a derived unit, the inverse second (1/s or s^{-1}).

F (1) The hexadecimal (HEX) symbol for binary 1111, decimal 15, the highest hexadecimal digit.
$F_{16} = 15_{10} = 17_8 = 1111_2$.
(2) Farad. Unit of capacitance, defined as the capacitance (C,5) of a capacitor (C,6) which requires a potential difference (PD) of 1 volt (V,1) to maintain a charge of 1 coulomb (C,9) on that capacitor. Named after Michael Faraday who discovered electromagnetic induction in 1831.

(3) Final. In computing this is a subscript, or lower case, symbol that designates *final.*

(4) Force. In the Système International d'Unités (SI) system of units, the unit of force is the newton (N). It is that for which, applied to a mass of one kilogram (kg), gives it an acceleration of one metre (m,2) per second (s) squared ($1\ N = 1\,kg\,m/s^2$).

FACE Field-Alterable Control Element. A chip (CHIP) employed in some microcomputer systems to permit the user to write microprograms. It allows for field microprogramming in low-volume applications and each system has a writeable control store, a logic unit and a display-and-debug unit.

FACT (1) Fast Access Current Text. An experimental electronic library, able to produce documentation facsimile (FAX) copies from microfiche via a telephone network.

(2) Fully Automated Cataloguing Technique. A library system that combines computer control of documents, along with storage on microfilm and a computer output microfilm (COM) facility.

F&E A type of video connector.

FAIRS Fully Automatic Information Retrieval System. A retrieval system that provides automatic searching of computer files containing information, in response to the definition of a particular user's needs.

FAMOS Floating-gate Avalanche-injection Metal-Oxide Semiconductor. A particular type of erasable programmable read-only memory (PROM) that erases by use of ultra-violet (UV) light and has an avalanche-transport mode. *See Figure F.1.* It has silicon-gate metal-oxide semiconductor (MOS) field-effect transistors (FET) that have no connection to the

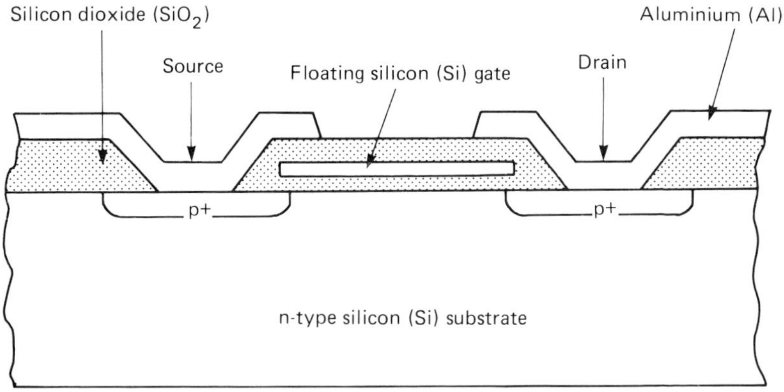

Figure F.1 A cross-sectional view of a floating-gate avalanche-injection metal-oxide semiconductor (FAMOS) structure

floating gate, and memory operation depends upon charge transport effected by avalanche injection from a source or a drain.

FAP Fortran Assembly Program. A procedurally orientated formula translator (FORTRAN) computer software system.

FAST Flexible Algebraic Scientific Translator. A device enabling a user to write a computer program in an algebraic format by translating complex statements into Beginner's All-purpose Symbolic Instruction Code (BASIC).

FAX FACSimile. A system used for the transmission of printed images. During operation the image is scanned by equipment at the transmitting station, transmitted and then reconstructed by the receiving station and finally duplicated as hard copy on paper.

FAXCOM FASCimile COMmunications. A service offered in North America by the TransCanada Telephone System. It allows for the transmission of written, drawn or typed graphics over a direct distance dialling (DDD) network.

FCC Federal Communications Commission. Appointed by the President of the USA, this is a board comprising seven commissioners who have the power to regulate all interstate and foreign electrical communication systems originating in the USA.

FD (1) Floppy Disk. A mass-storage device employing a soft flexible (floppy) diskette coated with magnetic material onto which information is recorded. The disk(ette) is driven so that it rotates inside a cardboard jacket and cutouts are provided for a moving head and index hole information.

(2) Full Duplex. The designation given to a method of operating a communications circuit so that each of its ends can simultaneously transmit and receive.

(3) ForwarD. A command in the Logo computer programming language that instructs a Turtle graphics robot to move forward a given distance.

FDC Floppy-Disk Controller. Controls the transfer of data to and from a floppy disk (FD,1). FDCs can interface both single-density and double-density disk drives and are able to interface up to four double-density drives. High-performance disk controllers can feature automatic seek, read and write operations, select lines for multiple heads, programmable loading and unloading of heads, control lines for phase-locked loops (PLLs) generally required for disk-drive circuitry and operate on a 5 volts (V,1) power supply unit (PSU).

FDF Fibre Distribution Frame. A type of optical fibre (OF) connector panel, whereby distribution and jumper cables, usually of a single fibre each, which can splice or connect optical transmitters to optical receivers, are able to be joined.

FDM Frequency-Division Multiplexing. A method of multiplexing in which more than one signal is transmitted on a channel (CHNL(a)) by dividing the available transmission frequency into narrower bands. A number of separate single channels are thus derived from splitting up a wider bandwidth into several narrower bandwidths.

FDMA Frequency-Division Multiple Access. A technique enabling communicating devices at various different locations to share a multipoint or broadcast channel (CHNL(a)). It allocates a different frequency to each user.

FDOS Floppy-Disk Operating System. (a) An operating system (OS) that has been developed to permit source programs written and edited at the system keyboard, to be stored on floppy disks (FD,1). The assembled programs are stored on the disk in binary form. A typical FDOS package can consist of dual drive floppy disk, interface, software, documentation and cable. (b) A combination of areas in the memory map (MAP,2) of a microcomputer using a control program for microcomputers (CP/M) disk operating system (DOS). The areas are the basic input/output system (BIOS) and the basic disk-operating system (BDOS).

FDX Full DupleX. The property of a communication system or particular piece of equipment to communicate simultaneously in two directions (*see* FD,2).

FE Format Effector. A specific control character used for the control of the layout or position (format) of printed or displayed information.

FEDLINK FEDeral Library and Information NetworK. A computer network in the USA providing an on-line or user direct cataloguing facility.

FEDREG FEDeral REGister. A store of information held on computer files providing details of US Federal Government regulations.

Fe_2O_3 Ferric Oxide. This red iron oxide is the magnetic constituent of nearly all magnetic recording tapes (MAGTAPE) used today. It is dispersed in fine particles to form a coating on the tapes.

FEP (1) Front-End Processor. The interface between a mainframe computer and the outside world. It is normally used to remove some of the necessary processing load from the computer.

(2) Fluorinated Ethylene Propylene. A 'Teflon' flourocarbon resin, which a registered trademark of the Du Pont Company.

It is melt extrudable flourocarbon resin.

FET Field-Effect Transistor. A semiconductor device with unipolar multielectrodes, where current flows through a narrow semiconductor material conducting channel (CHNL(h)) between two electrodes and is modulated (made more or less conductive) by means of an electric field applied at a third electrode. *See Figure F.2.* This three-terminal device consists of the channel which is a slice on n-type silicon (n-type) with a p-n junction (p-n) diffused into it. As FETs are controlled by voltage (V,2) rather than current (I), they exhibit low power requirements along with high impedance.

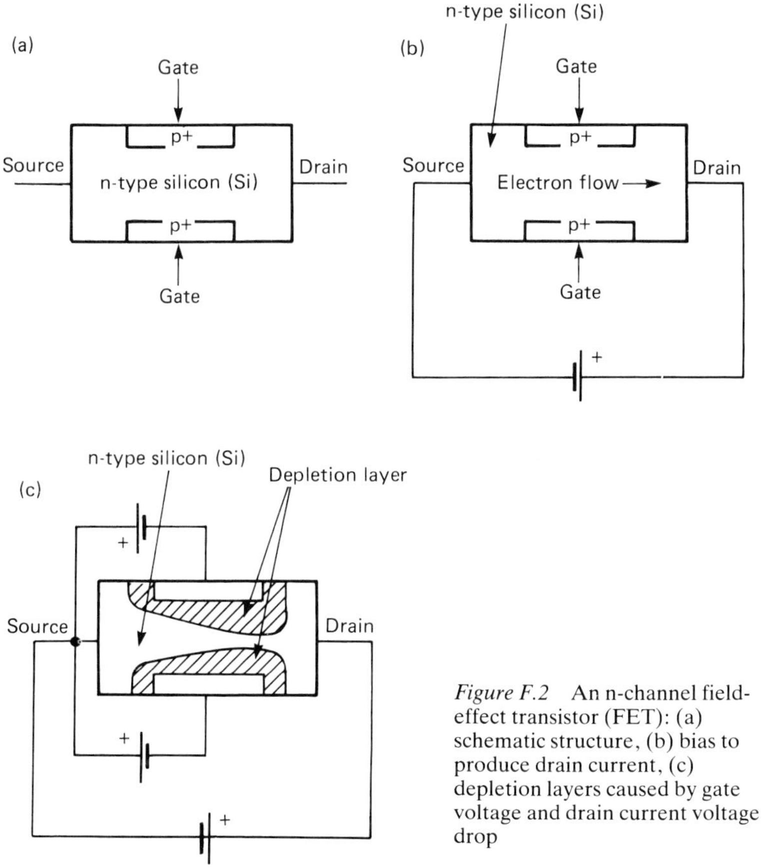

Figure F.2 An n-channel field-effect transistor (FET): (a) schematic structure, (b) bias to produce drain current, (c) depletion layers caused by gate voltage and drain current voltage drop

FETDIP FETlington Dual-In-line Package. The Fetlington is a trade marked product produced by Siliconix Inc. The family of products have been designed as single compact devices to replace Darlington configurations. The FETDIPs are auto-insertible 4-pin dual-in-line packages (DIP) and are JEDEC registered (*see* JEDEC). *See Figure F.3.*

Figure F.3 FETDIPS showing their auto-insertible 4-pin dual-in-line packages (*courtesy* Siliconix Inc.)

FEXT Far-End CROSSTalk. (a) Observation and measurement at a remote terminal in a telecommunications system of unwanted energy transfers between one circuit and another which allow the unwanted energy to travel in the same direction as the wanted signal. (b) In optical fibres (OF) the similar condition can be experienced as in (a). Measurement is at the receiving end of a given fibre when optical power is inserted into an adjacent fibre at its transmitting end.

FF (1) Form Feed. (a) A control character that provides regulation of the demands of a printer to feed forms. (b) A mechanical form-positioning system on a printing device that automatically locates and positions continuous stationery.
(2) Flip-Flop. A combination of two inter-connected active devices (i.e. a bistable) or two inter-connected logic gates. As long as power is available an FF circuit can be used to store one binary digit (BIT) of information permanently. *See Figure F.4.*

FFL Front Focal Length. A distance measured from the focal point, located in front of a lens in an optical system, to the nearest point on that lens.

FFT Fast Fourier Transform. Any realistic periodic signal can be represented as the sum of sine waves having various amplitudes and phases. This sum, although an infinite series, converges quickly. To include transients or non-periodic functions, the method for Fourier-series analysis is extended. In such a case the infinite series becomes a continuous function expressed as an integral, known as the 'Fourier transform'. To make available a method whereby calculations can be performed by

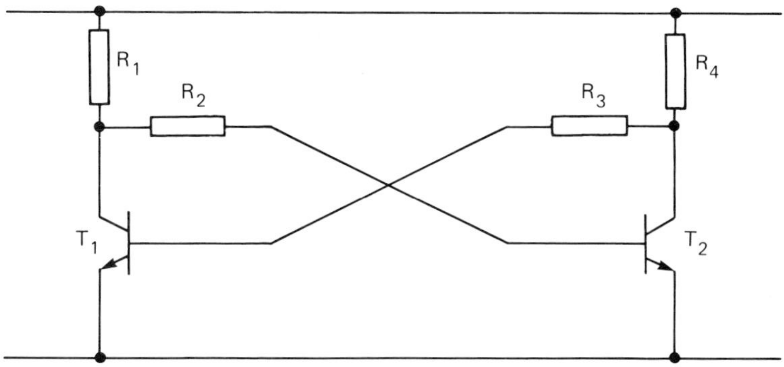

Figure F.4 A bistable circuit or flip-flop (FF,2)

computer, the Fourier transform (an infinite series) is provided with a finite range, known as the FFT.

FIFO First In, First Out. In memories, tables, stacks or lists, this is a method of storing and retrieving data or items. Information is deposited at one end, and removed from the other, enabling the oldest entry to be retrieved first.

FIGS FIGures Shift. A function initiated by a figure shift character and performed by a teletype that causes the machine to make a physical movement to the upper case in order that numbers, symbols and upper case characters can be printed.

FILO First In Last Out. Also known as a 'stack', this is a pop-up register which enables the most recent entry to be retrieved first.

FIMS Functionally-Identifiable Maintenance Systems. A system of block diagrams organized so that it can direct non-technical personnel to a replaceable unit. The unit can range from a computer (CMPT) to a printed circuit board (PCB), or to a simple repairable circuit (CCT).

FIP (1) Vacuum Fluorescent Indicator Panel. A specialized variation of a triode, having a directly heated cathode (filament), control grid, and segmented plate (anode) upon which a fluorescent material has been deposited. *See Figure F.5.* An electron flow from the filament is directed upon one or more of the anode segments, causing the phosphor on them to luminesce and emit light at a high brightness level. As the fluorescent material used is an efficient converter of electrical energy to light energy, FIPs operate at low voltages (V,2).

(2) Federal Information Processing standards. American National Standards Institute (ANSI) standards which have been adopted by the US Government.

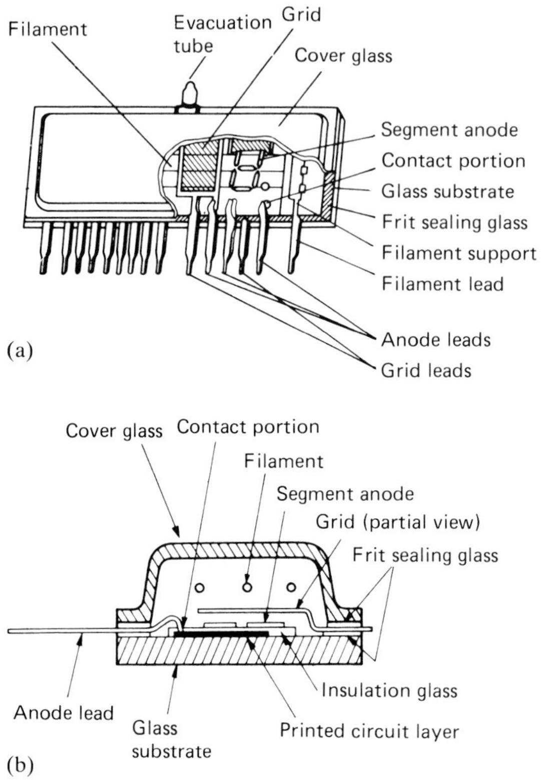

Figure F.5 Vacuum fluorescent indicator panels (FIP): (a) construction, (b) cross-section (*courtesy* NEC Electronics (UK) Ltd)

FIRST Fast Interactive Retrieval System Technology. An information retrieval technique that provides a system having the ability to allow the searching of computer files that contain particular items of data. The search is implemented under real time operator control.

FJCC Fall Joint Computer Conference. Originally organized by the American Federation of Information Processing Societies (AFIPS), these conferences that take place in the USA have now been superseded by the National Computer Conference (NCC).

FL Fusible Link. A fusible link has a fuse in the cell structure that is blown during programming when a higher than normal working current (I) is switched through the transistor and hence through the fuse in the emitter leg. Fusible link memory devices are completely non-volatile and once programmed cannot be erased. *See Figure F.6.* Continued improvements in fuse technology have resulted in four kinds of currently employed fuse materials, polycrystalline silicon, platinum silicide, nichrome and titanium tungsten.

FLOP FLOating Point. A notation to represent decimal numbers that is employed to help simplify arithmetic calculations by moving the decimal or binary point from left to right. 568 000 000 can be shown as 5.68, 8 as it is equal to 5.68×10^8.

FLOTOX FLOating gate Tunnel OXide. A principle employed in electrically-erasable programmable read-only memories (EEPROM) in which electrons tunnel to and from a floating gate. *See Figure E.2,* p. 81.

FM Frequency Modulation. A method of modifying a sine wave signal so that it carries information, as shown in *Figure F.7.* The sine wave or carrier wave (CW,1) has its frequency modified in accordance with other information required to be transmitted. The frequency function of the modulated wave may be continuous or discontinuous. In the discontinuous case two or more particular frequencies may each correspond to one significant condition.

FMS (1) File Management System. A collection of computer programs or an on-line executive program that provide the clear management of files. They allow for the creation, deletion or retrieval of files by name from a bulk storage device.
(2) Flexible Manufacturing System. Controlled by numerical control (NC) machines linked into a computer numerical control (CNC) system, an FMS makes use of robot-controlled transport of production work between machine stations.

F Number A camera iris calibration obtained through dividing the effective focal length by the effective iris diameter.

FO Fibre Optics. The activity and technology of the passive guidance of ultra-violet (UV), visible or infra-red light through predetermined paths along transparent flexible fibres. These fibres provide a path for a single beam of light or in multiples, such as the transportation of a complete image. The fibres are provided either singly or in a cable bundle.

FOC Fibre-Optic Communications. A communications system that provides for the carrying of a signal from one point to another by the use of optical fibres (OF) or fibre optics (FO).

FORMAC FORmula MAnipulation Compiler. A computer language which is an extension of formula translator (FORTRAN) but of a different type. It has been explicitly designed for algebraic operations.

FORTH A computer programming language. FORTH was

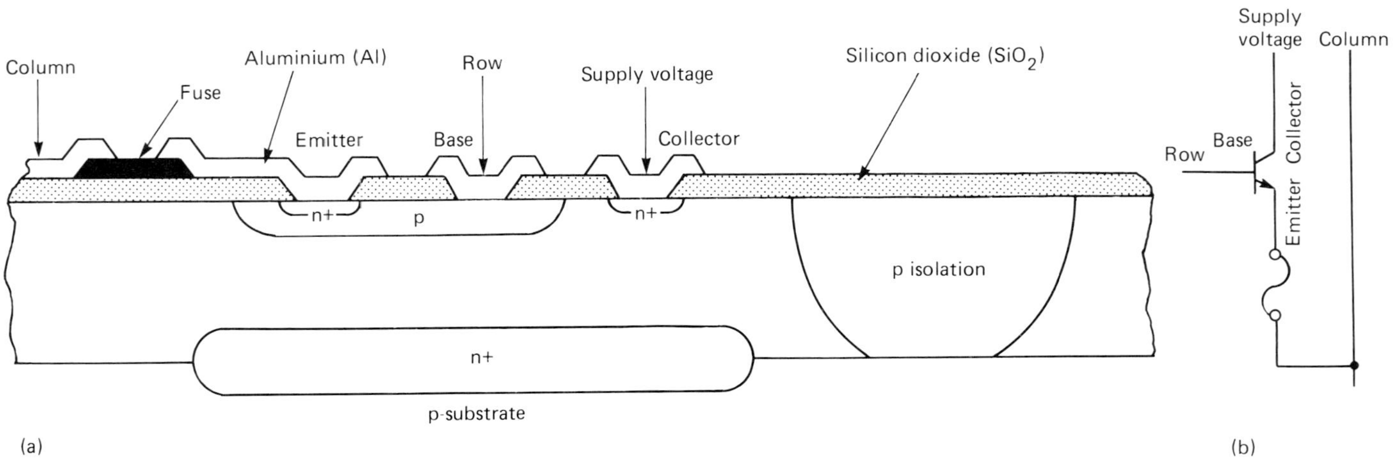

Figure F.6 (a) A cross-sectional view of the basic structure of a bipolar fusible link (FL) memory, (b) logic elements

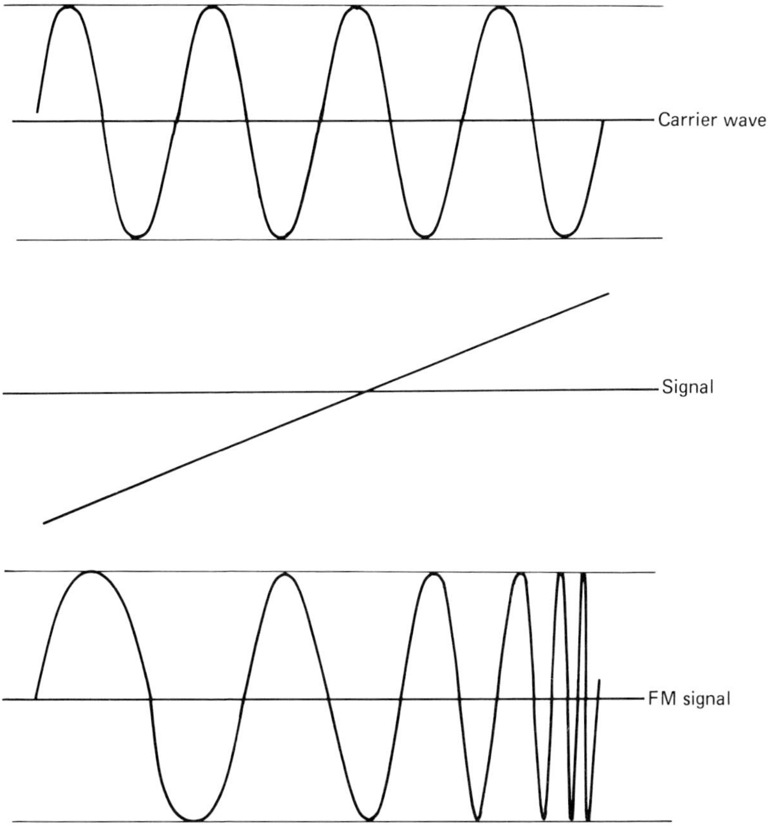

Figure F.7 Frequency modulation (FM)

invented in the 1970s by the astronomer Charles Moor to control one of the telescopes at Kitt Peak in the US. It also proved to be an alternative to the existing high-level computer languages.

FORTRAN FORmula TRANslator. A computer programming language originally conceived by International Business Machines (IBM) and developed for use on scientific analysis and mathematical problems. It is now extensively used for commercial and business applications. It is a widely used language in scientific environments.

FOSDIC Film Optical Sensing Device for Input to Computers. A device to enable a computer memory to receive information from a microfilm storage medium.

FOTS Fibre-Optic Transmission System. A system of transmission that uses small-diameter transparent fibres through which ultra-violet, visible or infra-red light is passed. *See* FO, fibre optics.

FPLA Field Programmable Logic Array. An arrangement of logic circuits in which the internal connections of the gates can be programmed by the use of a higher than normal working current (I), and can therefore be programmed by the user. Fusable links (FL) are used for programming the logic configuration and the high current is passed through the links to achieve the required logic design.

FPP Floating Point Package. A set of computer software routines that are necessary to perform floating point (FLOP) arithmetic.

FPROM Field Programmable Read-Only Memory. *See* FROM.

FPS Feet Per Second. A measure of speed, often used to indicate the rate at which a magnetic tape (MAGTAPE) passes a magnetic head.

FR Forced Release. A facility in a telecommunications system that enables a traffic carrying device or circuit to be released arbitrarily, quite independently of any release action taken by the caller or the called party.

FROM Field programmable Read-Only Memory. A read-only memory (ROM) that is programmed by fusing or burning out particular critical interconnections at specific storage locations. (*See* FL, fusible links.) The programming can be completed by the customer using special equipment.

FS (1) File Separator. An indicator or flag employed to show the boundary between one collection of related records treated as a whole unit (file) and another.

(2) Final Selector. A telecommunications switching unit that has one inlet and provides direct access to a number of subscribers' lines via its outlets. Operation can be by external or internal controls.

(3) Field Separator. A control character employed to indicate or flag the boundaries of a group of contiguous locations (or columns on a card) which contain a particular item of data or a group of related pieces of information.

FSD Full-Scale Deflection. (a) A term normally applied to an analogue meter, in which the indicating pointer of the device reaches the maximum (MAX) value for which the instrument has been scaled. (b) The absolute maximum (MAX) value or full scale value of a measured amount of whatever specific quantity for which a particular measuring instrument has been calibrated.

FSK Frequency-Shift Keying. A form of frequency modulation (FM) where the frequency is made to vary without interrupting the carrier at the instant limiting intervals of modulation (i.e. the significant conditions of the modulating digital signal), between a set of predetermined discrete values called the 'mark' and the 'space' frequencies. Generally FSK is used in telecommunications for lower speed devices such as teleprinters.

FSL Formal Semantics Language. A language used for computer-orientated languages that is employed as a machine-independent language. It intergenerates compilers for a specific machine and contains formal semantics.

FSP Frequency-Shift Pulsing. A technique of signalling, used in telecommunications, which makes use of a frequency shift between 1070 Hz and 1270 Hz. This particular type of signalling is used with narrow-band systems, such as teletypewriter (TTY) switching networks.

FSS Fixed Satellite Services. FSS includes satellites such as Intelsat I (*see* INTELSAT), and uses low power relays for telephony and television (TV(a)) distribution. Large antennas are often required to receive these signals.

FST Flat Square Tube. Refers to a television (TV(c)) colour picture tube (CPT) in which the picture area is square (rather than the 4:3 rectangular ratio) and has a flat rather than domed surface.

FSTV Flat-Square TeleVision. Any television (TV(c)), often pocket-sized, which uses a tube flatter than usual. There are two types, a cathode-ray tube (CRT) deflected through 90 degrees and a liquid crystal display (LCD).

FSU Final Signal Unit. A group of information and check binary digits (BITs), normally subject to a single error detecting process, that are used in the common-channel signalling (CCS) or multi-unit messages. The International Telegraph and Telephone Consultative Committee (CCITT) have designated a signal unit system No. 6 which consists of 28 bits, the last being used for error-detection purposes.

FTET Full Time Equivalent Terminals. The access to a computer system of a particular terminal can be quantisized by using FTET. It allows for the number of hours that the terminal is available for direct on-line access.

FTP File Transfer Protocol. Permitting transfer between different computers, FTP is a formalization of the methods by which the two parties communicate.

FTS Federal Telecommunications System. A leased telephone network in the USA shared by federal government agencies.

FX (1) Foreign eXchange. Provides connection between a customer and a remote telephone exchange, giving the equivalent of local service from the distant exchange.
(2) FiXed area. A portion of a computer disk where core image programs can be stored and protected.

G

g (1) Gram. The unit of mass in the centimetre, gram, second (CGS) system of units, it is defined as one thousandth of a kilogram (kg).
(2) Grid. An electrode situated between the cathode (C,10) and the anode (A,3) of an electron tube. The main electron stream of the tube passes through its intersections.
(3) Gravitational constant. The acceleration of gravity varies over the Earth's surface because the Earth is not a sphere. However, for most practical purposes it is inconvenient to have units of measurement which vary from place to place on the Earth's surface. Therefore a standard gravity (gravitational constant) is now accepted internationally. The value denoted by g, is $9.80665\,m/s^2$. *See* metre (M,2) and second (s).

G (1) Giga. A decimal prefix commonly used in association with a base unit in the Système International d'Unités (SI) system of units, indicating a multiplication factor of 10^9 of that unit. For example, one gigahertz (GHz) is equal to 1 000 000 000 hertz (Hz).
(2) Gauss. In the centimetre, gram, second (CGS) system of units the gauss is the unit of magnetic flux density. Now superseded by the tesla (T), one gauss equals 10^{-4} tesla.
(3) Generate. An output (O,2) from an adder in a microprocessor (MP) that has been designed for the connection to a carry look ahead circuit.
(4) Ground. Also known as 'earth'. An intentional or accidental connection made to earth, or a large conducting body. The earth connection provides a reference potential level and is also employed to carry earth currents in communication circuits.

G^3 Gadolinium Gallium Garnet. A nonmagnetic substance, grown as single crystals. When cut into wafers, it is used as a substrate for solid-state magnetic circuits.

GaAs GAllium ArSenide. A semiconductor material that can be used for high-speed applications. It is difficult to work with and its main uses have been for microwave devices, but lately it has been employed for metal semiconductor field-effect transistors (MESFETs) used in high-speed logic circuits.

GALPAT GALloping PATtern memory. A microprocessor

(MP) test technique for generating successive patterns.

GAP General Assembly Program. Allows programs to be written in symbolic code rather than in the absolute machine language of a computer.

GB Grid Base. The span or range of grid voltage (V,2) between zero and the voltage that provides a cut off of anode current in a gas-filled electron tube. It thus provides control of the current (I) flow between the two other electrodes.

Gbps GigaBits Per Second. An information transfer or signalling rate of 10^9 binary digits (BITs) per second.

GDF Group Distribution Frame. In a frequency division multiplexing (FDM) system, GDF is a frame that provides for the interconnection apparatus used in the twelve transmission channels (CHNL(a)) assembled in frequency band 60–108 kHz.

GDO Grid DipOscillator. Used for the measurement of radio frequency (RF), this portable instrument employs a grid-current meter and an oscillator. Resonance in an antenna is detected when the meter indicates a sharp dip.

GDT Graphic Display Terminal. A visual display unit (VDU) that permits the user or operator to display graphics information.

Ge Germanium. A tetravalent crystalline element, atomic number 32. It is widely used in the manufacture of semiconductor devices.

GE (1) Greater than/Equal to. A comparison of two values that states whether one value is greater than or equal to another. Normally it is used in order to allow another process or calculation to be carried out.

The symbol for GE is $\geqslant$.

(2) Gateway Exchange. A telephone exchange that allows the connection of international exchanges with a national network.

GEM Graphics Environment Manager. GEM is an operating environment, developed by Digital Research Inc for IBM PCs and compatibles. GEM was really the first of what has become known as windows, icons, mouse and pull down menus (WIMP) to be released. It is designed to offer a friendly interface between the operating system (OS) and the user by employing pictures to depict files of different types and allowing them to be chosen and executed by the use of an on-screen pointer.

GEORGE GEneral ORGanizational Environment. An organized collection of techniques and procedures for operating a computer, that resides on a disk, employed with ICL computers.

GHz GigaHertZ. A frequency of one thousand million hertz (Hz).

GI Graded Index. Characteristic of an optical fibre (OF) which has a refractive index that varies and is lower at distances radially away from its axis. This causes light rays to be continually refocused by refraction in the core. *See Figure G.1.* It bends the rays inwards and allows them to travel faster in the lower index of refraction region. A GI fibre provides high band-width capabilities.

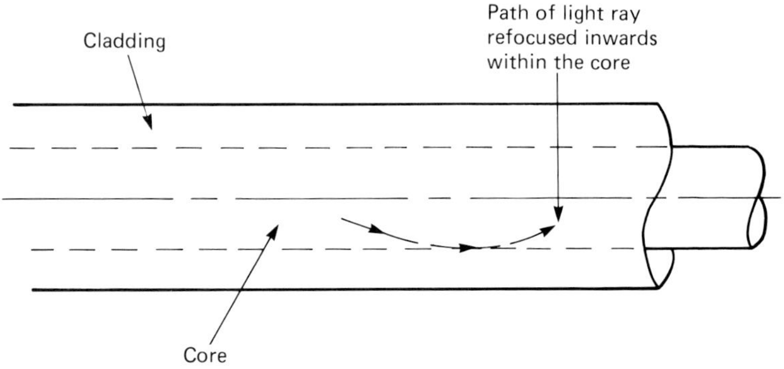

Figure G.1 Graded-index (GI) optical fibre (OF)

GIGO Garbage In, Garbage Out. Produced as a result of undesirable input information or data, GIGO is unwanted and meaningless information carried along in the storage of a computer.

GIRL Generalized Information-Retrieval Language. A computer language employed in the searching of stores of data held on computer files. It has been developed by the US Defense Nuclear Agency.

GJP Graphic Job Processor. A feature of a computer program for job control, that allows job control statements and other associated information to be entered via a remote visual display unit (VDU).

GKS Graphical Kernel System. A standard defined in ISO GKS 7.2 for computer systems. Comprising a set of basic functions for graphical input/output (I/O) on vector and raster devices, GKS makes for portability of application programs between installations, savings in programming effort and simplified viewing of a variety of display devices. It can support several work stations in parallel.

GM Group Mark. A special character used in a write instruction to designate the beginning or end of a group of words, characters or records in storage. It enables the group so identified to be processed and addressed as a unit.

GND GrouND. Usually connected to earth for safety, the GND is a point on a circuit at zero volts (V,1) potential. All points on a circuit at this zero potential can be said to be connected by a common ground.

GOS Grade Of Service. A measure of service provided by a telephone exchange during a period of an hour in which the average traffic intensity is at maximum (MAX). It is usually expressed as the ratio of telephone calls liable to failure at the first attempt due to equipment limitations.

GOSUB GO to SUBroutine. A statement in beginner' all-purpose symbolic instruction code (BASIC). It is used to avoid the repetition of writing a section of the program.

In practice the subroutine is used for parts of a program that may be repeated a number of times. It is kept apart from the main program. After it has been used, execution will return to the correct point in the program, i.e. the first line after the GOSUB statement.

GPC General-Purpose Computer. A generic term for computers (CMPTs) that can be used for a number of purposes, rather than dedicated to a specific function or functions.

GPIA General-Purpose Interface Adaptor. Able to interface between an IEEE-488 standard instrument bus and a microprocessor unit (MPU), the GPIA enables many instruments to be interconnected and thus remotely automatically controlled or programmed.

GPIB General-Purpose Interface Bus. A bus that connects instruments, including the connectors, the signal lines and the voltage level on lines, the functions used by the instruments, interfaces to exchange data and control messages and protocols. The mechanical specification defines the cable and connectors and the electrical specification defines the voltage (V,2) and current (I) values required to be compatible with transistor transistor logic (TTL).

GPO General Post Office. UK organization now split into two parts – the Post Office (PO), which handles postal service, and British Telecom (BT), which handles telephones and telecommunications.

GPR General Purpose Rubber. A material used for cable insulation.

GPSDW General-Purpose Scientific Document Writer. Capable of producing complex hard copies for scientific use, this computer output (O,2) device can deal with mathematics and graphs.

GPSS General-PurpoSe Simulation program. A high level computer programming language. Each natural language instruction corresponds to a number of machine code instructions. It permits the use of notations with which the user is already familiar.

GRO Gamma-Ray Observatory. A planned Earth-orbiting observatory, which has been designed to investigate gamma-rays from stars and galaxies, and is due for launch by the space shuttle.

GRP Group Reference Pilot. A number of frequency waves transmitted to help maintain and adjust a system. It is applied where telephone channels (CHNL(a)) are assembled into groups, accompanying the group until they are separated into their individual channels.

GRTS General electric Remote Terminal System. A synchronous protocol developed by General Electric in the US.

Gs Gauss. In the centimetre, gram, second (CGS) system of units the gauss is the unit of magnetic flux density. Now superseded by the tesla (T), one gauss equals 10^{-4} tesla.

GS Group Separator. A special character used to designate the beginning or end of a group of words, characters or records in a store. It enables the group to be identified and be processed and addressed as a unit.

GSC Group Switching Centre. A term used in the UK for a telephone exchange that connects a number of local exchanges to traffic routes between exchanges more than 24 km (15 miles) apart.

GSI Graphic Structure Input. Permitting the direct searching of a store of information held on a computer file, GSI methods are used in connection with scientific material and data.

GSM Group Special Mobile. A committee set up by the conference of European Postal and Telecommunications Administrations (CEPT) directing the establishment of a Pan-European cellular radio system.

GSW Galvanized Steel Wire. Drawn steel wire protected from corrosion by a layer of zinc (galvanizing). GSW is commonly used in the production of single-wire armour (SWA) cable.

GT Greater Than. A comparison of two values that states whether one value is greater than another. Normally it is used to allow another process or calculation to be carried out. The symbol for GT is >.

G/T Gain over noise-Temperature. A ratio of gain over noise-temperature, in a receiver (RX) system, relating the gain of an antenna to the system noise and the surrounding noise.

GTH Gas-Tight High-pressure. A method of stacking printed circuit boards (PCBs) and providing efficient connection without the need for goldplating. The spring contacts dig relatively deeply into the softer PCB pad, to provide a gas-tight seal around the metal contact. The GTH principal has also been incorporated into the design of a printed circuit board (PCB) mounted, zero insertion force (ZIF) termination for flat or descrete solid standed conductors. A simple example of each is shown in *Figure G.2.*

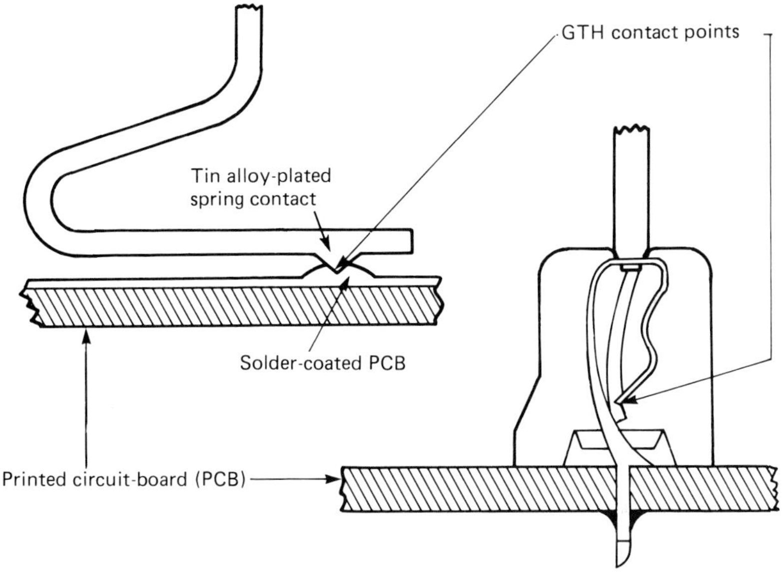

Figure G.2 Two examples of gas-tight high-pressure (GTH) contacts

GTN Government Telecommunications Network. A network that links more than 400 government offices throughout Britain by means of tandem exchange and private circuits rented from British Telecom (BT).

GTO Gate Turn-Off. The name given to a type of device utilizing a thyristor switch which is activated by applying a negative pulse to its gate. The GTO combines the blocking voltage (V,2) and overcurrent capability of the thyristor with the ease of gate drive and fast switching of the transistor.

GUTS Gothenburg University Terminal System. Designed and implemented by the University of Gothenburg, GUTS is a particular type of interactive system available on International Business Machines' (IBM) equipment and machines.

h Hecto. A decimal prefix used in association with a base unit in the Système International d'Unités (SI) system of units indicating a multiplication factor of 10^2 of that unit. For example, one hectometre hm is equal to 100 metres (m,2).

H (1) Henry. The unit of inductance and permeance in the Système International d'Unités (SI) system of units.

(2) High. The most significant half (that containing the most significant digits (MSD)) of a microcomputer memory location or pointer. Normally, the 0–7 portion of a 16 binary digit (BIT) word.

(3) Quadripole. A prefix used in relation to quadripole network. It has a pair of inputs and a pair of output (O,2) terminals. *See Figure H.1.*

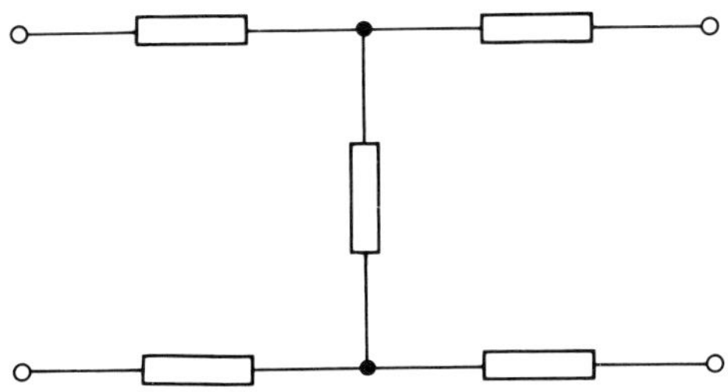

Figure H.1 A quadripole (H) network

HA Half Add. A computer instruction that carries out binary digit (BIT)-by-binary digit half additions. The operation is by a half-adder device where two inputs are received and two outputs (O,2) are produced representing the sum and the carry.

HAL Harwell Automated Library. A computerized system run by the Atomic Energy Authority at Harwell that provides for the loan of items of scientific data from the Authority's library.

HAM A slang expression used to describe an amateur radio operator.

HAMT Human-Aided Machine Translation. A computerized system capable of automatically translating text from one spoken language to another. It requires human intervention only to resolve ambiguities of syntax and semantics as well as problems arising from non-literal usage.

HASP Houston Automatic Spooling Priority system. Providing spooling and priority functions for a computer operating system (OS) made by International Business Machines (IBM), this system has a number of modified versions.

HC (1) Heat Capacity. A measure of the amount of heat that can be absorbed by a component. In the case of a diode (D,5) during operation, it would be the heat absorbed during pulses so short that little heat flows out of the diode into its heat sink during the pulse.

(2) Heat Coil. A thermally-operated protection device connected between a transmission line and a telephone exchange. It operates at temperatures that are too low for other protection devices such as fuses.

HD Half-Duplex. (a) Generally in computer terminology, a circuit or system that can transmit in either direction, but not both directions at once. (b) A circuit designed for duplex (DX) operation, but limited by the nature of the terminal equipment to use in alternate directions only.

HDB High Density Bipolar. A version of coding used in digital (D,4) transmission. Long strings of zeros are replaced by coded groups containing timing information.

HDLC High-level Data Link Control. A formalized method of communication in which information is considered as a stream of binary digits (BITs), allowing data transparency via bit stuffing.

HDX Half-Duplex. *See* HD.

HELP The HELP mode. A particular method of operation of a computer, terminal or processor. The user is guided by the HELP mode which indicates instructions on a visual display unit (VDU). The mode enables the user, with no programming experience, to perform functions or obtain information by using single keystrokes.

HEMT High Electron Mobility Transistor. A two-dimensional electron gas (2 DEG) at the heterojunction interface of a microwave (μW) transistor which can be used to produce a specific type of field-effect transistor (FET) known as a HEMT. Such devices exhibit vast improvements in maximum (MAX) frequency of operation and minimum (MIN) noise figures, possibly the lowest noise figures for microwave (μW) applications.

HEPR Hard grade Ethylene Propylene Rubber. Having properties similar to those of crosslinked polyethylene (XLPE), HEPR is a hard grade option of ethylene propylene rubber (EPR) cable insulation.

HEX HEXadecimal. A system of number notation to the base 16. The most commonly used set of digits representing the hexadecimal notation system are 0 to 9 and then A, B, C, D, E and F. Each hexadecimal digit can thus be represented by a group of four digits (BITs), which is called a Tetrad. There are some other systems not so widely used that employ letters other than A-F for the last six digits (DGTs).

HEXFET HEXagonal mosFET. A hexagonal cell structured power metal-oxide silicon field-effect transistor (MOSFET) produced by International Rectifier.

HF High-Frequency. In radio systems, the HF bands lie between 1·6 and 30 MHz, and are popular because relatively long distances can be covered at low cost, under relatively light traffic conditions.

Hg Mercury. A metallic element, atomic number 80, atomic weight 200.59, specific gravity 13.6, melting point −39°C and boiling point 357°C. It is used in thyratrons, arc rectifiers, tilt switches and to 'wet' the contacts of various switches.

HICLASS HIerarchical CLASSification. One in which terms are provided with a designation and arranged in a hierarchical order. Items are split into initial sets, and then into successively finer sets, e.g. the Dewey decimal classification (DDS) system.

HI-FI HIgh-FIdelity. A system of sound reproduction that corresponds very closely to the original. Various national and international standards have been laid down covering HI-FI.

HiNIL HIgh-Noise-Immunity Logic. A specific design of logic circuits for use in applications where transients are likely to be present producing a noisy environment. They can protect complementary metal-oxide semiconductor (CMOS) circuits from transients and static electricity, and are compatible with a number of analogue circuits.

HIRAM Highest position Random-Access Memory. The highest address at which program data can be held in the random-access memory (RAM) of a microcomputer. Also known as RAMTOP.

HISAM Hierarchical Indexed Sequential Access Method. A technique of organizing a store of information entered on a computer file, that can be held on a disk.

HJ Hetero Junction. Refers to devices that have several deposited layers and hence a number of p-n junctions (p-n).

HLL High-Level Language. A computer programming language resembling natural language using symbols and command statements that an operator can read, but is not English. Each statement represents a powerful series of computer instructions. Examples are: Beginners all-purpose symbolic instruction code (BASIC), common business-orientated language (COBOL) and formula translator (FORTRAN).

HMI Halogen Metallide Iodide. Very efficient lighting, which, however, can cause strobing effects on film and television (TV).

HMOS High-speed Metal-Oxide Semiconductor. A metal-oxide semiconductor (MOS) technology that employs high-scaling techniques to increase density (D,3) and performance beyond that of n-channel semiconductor (NMOS) and p-channel metal-oxide semiconductor (PMOS) technologies.

HNIL High-Noise Immunity Logic. *See* HiNIL.

HOFR Heat resistant, Oil resistant, Flame Retardent. Refers to a standard of cable suitable for use in hot situations such as night storage heaters. Also for situations involving contact with oil, grease and varnish and for power and lighting services, reeling and trailing applications for exterior equipment such as mobile cranes and conveyors.

HOL High-Order Language. The same as high-level language (*see* HLL).

HP-IB Hewlett-Packard Interface Bus. An implementation of IEEE-488 (*See* IEEE-488) standard for digital (D,4) interfacing of programmable instrumentation made by Hewlett-Packard.

HRC High Rupturing Capacity. Usually refers to a type of cartridge fuse which has a high performance. Its minimum rupturing capacity is above 16 500 amps (A) and it is designed to cut off rapidly.

HSM High-Speed Memory. A computer peripheral device that is capable of retrieving information at a comparatively faster rate than other memory units connected to a computer system.

HSP High-Speed Printer. A printer that is able to operate on-line and has a speed of operation compatible with data processing (DP) and computer speeds.

HSR High-Speed Reader. A device that converts information from one storage medium into another. In general a card reader operating at a speed more compatible with the speed of computation and data processing (DP) so that it may operate on-line.

HSYNC Horizontal SYNChronization. A signal employed in a television (TV(c)) set to determine the horizontal position of the picture image on the screen.

HT (1) High Tension. A high-voltage supply required for anodes and screen grids of electron tubes. Usually in the range of 60–250 volts (V,1).

(2) Horizontal Tabulation. A specific control character that causes the position of print of a printer to be pushed forward to the next allocated position in the same line of print.

HTL High Threshold Logic. A type of microcomputer logic employed in industrial environments, where the electrical noise level can be quite high. Digital systems operated in such environments must have a high threshold in order to ignore these noise spikes, and thus not provide false operation. The

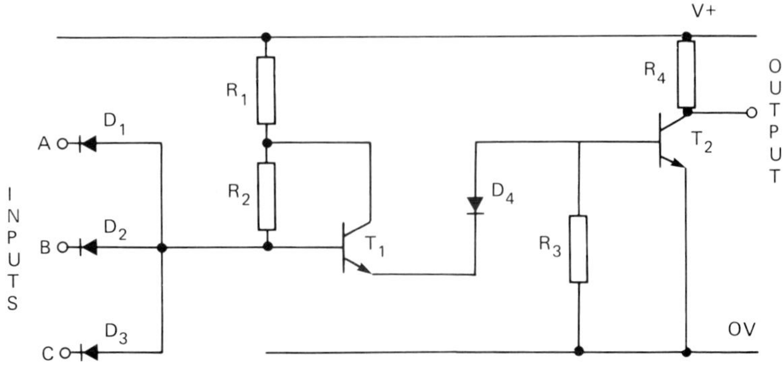

Figure H.2 A typical high-threshold logic (HTL) gate

HTL gate is a development of diode transistor logic (DTL). A typical HTL gate is shown in *Figure H.2.*

HUD Head-Up Display. An arrangement of equipment that projects images into a pilot's line of sight. It uses a cathode-ray tube (CRT) to provide comprehensive information on height, speed, heading and altitude.

HV High Voltage. A term used in reports and documents which generally refers to a voltage of more than 650 volts (V,1).

HVDC High-Voltage Direct Current. A direct current (DC) with a voltage in excess of 650 volts (V,1).

HVMOS High Voltage Metal-Oxide Semiconductor. A high voltage (V,2) technology based on the double-diffused metal-oxide semiconductor (DMOS) process which provides high voltage switches (both linear and non-linear) together with a polysilicon gate complementary metal-oxide semiconductor (CMOS) capability on the same chip (*see* CHIP,1). The areas of application include relay replacement circuits, semiconductor crosspoints, interface between high voltage and peripheral electronics, flat panel display drives and white goods circuits.

HVPS High-Voltage Power Supply. A form of power-supply unit (PSU) that can provide a high-voltage (HV) supply output (O,2) from a lower-voltage input. A typical HVPS is found within a domestic television (TV(c)) receiver and is used to provide thc cathode-ray tube (CRT) element voltages (V,2).

HYPERNET HYPER NETwork. A generic term used to describe a number of computers interconnected to form a network capable of transmitting information at megabaud rates.

Hz HertZ. The unit of frequency in the Système International d'Unités (SI) system of units equal to the number of cycles per second (C/S). For example, a frequency that has 3 complete cycles in a second is said to be at 3 hertz.

I

I Current. The rate of flow of electricity past a given point. The unit of measurement is the ampere (A).

IAC Interference Absorption Circuit. A circuit normally found in receivers (RX) within cars. It suppresses external interference as well as that from the car.

IACS International Annealed Copper Standard. A standard of purity and conductivity laid down by the International Electrotechnical Commission (IEC). Annealed copper is said to have a conductivity of 100% IACS.

IAD Initiation Area Discriminator. A type of specialized cathode-ray tube (CRT).

IAEA International Atomic Energy Agency. Formed on the 26th October 1956, the IAEA is an autonomous intergovernment organization dedicated to increasing the contribution of atomic energy to the world's peace and ensuring that the agency's assistance is not used for military purposes. Based in Vienna it has over 100 members.

IAL International Algebraic Language. A computer language which was the predecessor of algebraic-orientated language (ALGOL).

IAM Initial Address Message. The first signal message transmitted in a telecommunications call set-up process, in common-channel signalling (CCS). It contains all the information necessary to identify the called party or the route to be taken.

IAR Instruction Address Register. A device that temporarily stores the location in a memory (address) of the next instruction to be executed.

IB Infinite Baffle. A loudspeaker mounting in which there is no air path between the front and the back of the loudspeaker. An IB necessitates an air-sealed enclosure.

IBA Independent Broadcasting Authority. So called because of its independence from any control or finance provided from licence fees (as compared to the BBC). The IBA controls all UK commercial television and commercial sound radio.

IBIS International Book Information Service. Providing information for Prestel, this service's major activity is sending information concerning recent publications to subscribers. The subscribers' particular interests are held on a computer file.

IBM International Business Machines. The largest manufacturer of computers in the world.

IBM PC International Business Machines Personal Computer. The personal computer (PC,2) produced by IBM, which is used throughout the world. It has spawned many compatible systems produced by other manufacturers.

IBM PC AT International Business Machines Personal Computer Advanced Technology. Introduced in the autumn of 1985, the IBM PC AT utilized the then new Intel 80286 microprocessor chips (CHIP) to form the basis of a range of fast microcomputers. Able to access in excess of three megabytes of memory when using the Xenix operating system (OS), it was the first serious IBM desk top system aimed at the true multi-user market. In its original form, it came with 512 Kb of memory and a 20 Mb fast Winchester Disk plus a new 1.2 Mb floppy disk drive. It was able to run most of the vast range of software available for the established IBM PC XT.

IBM PC ATE International Business Machines Personal Computer Advanced Technology Enhanced. A version of the IBM PC Advanced Technology (see IBM PC AT) which employs a different systems board.

IBM PC ATX International Business Machines Personal Computer Advanced Technology eXpanded. A version of the IBM PC Advanced Technology (see IBM PC AT) which employes a different systems board.

IBM PC XT International Business Machines Personal Computer eXpanded Technology. Announced in the spring of 1983, the IBM PC XT heralded IBM's entry into the high capacity, high speed world of Winchester Disks. Based on the successful IBM PC, it utilized the same 8088 microprocessor chip (CHIP), manufactured by Intel. It had a 10 megabyte, high speed Winchester Disk plus a single 360 Kb floppy disk (FD,1) and allowed memory expansion to 640 Kb. It also had an RS-232 (see RS-232C) serial communications port as standard.

IC (1) Integrated Circuit. (a) A single-package complete circuit (CCT), able to perform at least one, sometimes many, complete functions. This interconnected array of components is fabricated from a single crystal of semiconductor material by etching, doping and diffusion. Generally, ICs are known by their circuit's complexity as follows: small-scale integration (SSI), medium-scale integration (MSI), large-scale integration (LSI), very-large-scale integration (VLSI) and extra-large-scale integration (ELSI). (b) IC represents integrated circuits on parts lists and schematic diagrams.
(2) Instruction Counter. A multibit device which temporarily stores information that monitors the location in a memory (address) of the current instruction and is used to input information in the memory-address register (MAR).

ICA Integrated Communications Adaptor. An integrated adaptor that permits connection between multiple communication lines and a processing unit.

ICAND MultiplICAND. One quantity or amount that is multiplied by another quantity or amount.

ICC International Computation Centre. The computer centre's services are available to member nations of its sponsoring organization UNESCO.

ICE (1) In Circuit Emulator. A device that actually replaces a microprocessor (MP) by connecting directly to the circuit board (in circuit). The device has a hardware/software capability for the real time (RT,1) input/output (I/O) debugging of chips (CHIPs). The emulated microprocessor can be stopped, its memory locations can be examined and modifications as well as other alterations can be carried out.
(2) Interface Cancellation Equipment. An electronic counter-counter measure (ECCM) system for all types of very-high-frequency (VHF) combat radio. The system can upgrade current equipment to meet electronic counter measures (ECMs) used against battlefield VHF communications.
(3) Insertion Communication Equipment. A British Broadcasting Corporation (BBC) communication system that employs a line in field blanking.

ICIREPAT International Co-operation Information Retrieval among Examining PATent offices. This Geneva-based organization promotes international co-operation between numerous national patent offices and is concerned with the documentation and retrieval systems run by the various offices.

ICL International Computers Limited. A major manufacturer of computers in the UK.

ICM InComing Message (tape). Refers to the tape in a telephone answering machine that records callers' messages.

ICS Integrated Communication System. A system incorporating two or more autonomous communication systems combined into one single system, so that none of the original autonomy remains.

ICT In Circuit Test. The examination of discrete components and integrated circuits (IC) as part of a complete circuit (CCT). This can be done statistically in a non-energized state and dynamically in an energized state. The term is used generally in association with tests via automatic test equipment (ATE).

ICU Instruction Control Unit. Typically this device contains a read-only memory (ROM) storing the microinstructions and address control logic for branching. This part of a computer enables instructions to be retrieved in the correct order, interpreted and applied.

ICW Interrupted Continuous Wave. A type of controlled

disturbance propagated through space (wave), used for radio telegraphy where the audio-frequency (AF) wave is keyed on and off.

ID (1) Insulation Displacement. The technology employed in insulation displacement connectors. (*See* IDC.)
(2) IDentification. A computer, data processor (DP) or telecommunications label that can consist of a code number or code name which serves to uniquely identify a record, block, file or other unit of information or data.

IDC Insulation Displacement Connector. A type of connector and connector system that provides a self-stripping contact. The contacts are normally in the shape of a 'U' formed by two parallel cantilevered beams or legs with sharpened tips. The majority of flat cables for use with IDC systems are manufactured on 0.050-inch centres. A complete flat cable with as many as 64 conductors can be terminated in one process.

IDDD International Direct Distance Dialling. An automatic system of calling telephones beyond the local area, enabling subscribers to make international calls. *See* DDD.

IDF Intermediate Distribution Frame. A unit that provides for the interconnection, termination or cross-connection of cables and wires from units of telephone exchange equipment.

IDN Integrated Digital Network. A network using both digital transmission and digital switching.

IDP (1) Integrated Data Processing. (a) A computer system processing concept in which the processing of data is organized, correlated and systematic, embracing all the data processing (D,1) requirements in a particular area of interest. (b) Treating data processing requirements as a whole, this system tries to reduce or eliminate the duplication of data entry or processing steps. (c) A system where all procedures are linked by a computer employed for the processing of data.
(2) InterDigit Pause. An interval of time experienced when telephone dialling, between the end of one set of pulses which represent a digit (DGT) and the start of the next.

IDPM Institute of Data Processing Management. A professional body based in the UK, formed from the Institute of Data Processing and the Data Processing Management Association.

IDU IDle signal Unit. A group of information and check binary digits (BITs) used in common-channel signalling (CCS) during a period when no signal messages are free for transmission.

IEC International Electrotechnical Commission. Probably best known for the formulation of the MKS (metre, kilogram, second) system of fundamental units, the IEC, which is based in Geneva, has produced much standardization in the electrical field since its formation in 1906. It recommends standards for electrically-operated equipment, working with the International Standards Organisation (ISO) in this particular area.

IECQA International Electrotechnical Commission Qualification Assessment. Operating under the authority of the IEC, the IECQA defines and implements quality assessment procedures for electronics components, which are not mandatory for members of the IEC.

IEE Institution of Electrical Engineers. A UK professional body that promotes the general advancement of electrical/electronics science and engineering and their application: facilitates the exchange of information and ideas on these subjects; and financially aids the promotion of invention and research. It is also concerned with the standardization of electrical/electronics equipment and offers to the British Standards Institution (BSI) advice for determining national standards.

IEEE Institute of Electrical and Electronics Engineers (US). A professional body concerned with the standardization of electronic equipment. It also determines standards on behalf of the American National Standards Institute (ANSI).

IEEE-488 A standard of the Institute of Electrical and Electronics Engineers (IEEE) that is widely used to specify a microprocessor (MP) parallel bus ideal for high-speed module-to-module communication as well as for system-to-system communication. *See* GPIB.

IEEE-583 Usually called CAMAC, this is a standard laid down by the Institute of Electrical and Electronics Engineers (IEEE). It is a hardware/software specification that was originally developed for the nuclear industry but is currently widely adopted by many other industries. *See* CAMAC.

IEEIE Institution of Electrical and Electronics Incorporated Engineers. A UK professional body that provides for electrical and electronics engineers and technicians a recognized engineering qualification of a high level.

IERE Institution of Electronic and Radio Engineers. A UK professional body that promotes the general advancement of the science and practice of radio and electronics engineering, and facilitates the exchange of ideas and information on these and other branches of engineering and their applications.

IEV International Electrotechnical Vocabulary. A standardized set of electrical terms, devised by the International Electrotechnical Commission (IEC).

IF (1) Intermediate Frequency. A carrier frequency to which the modulation of all received signals is transferred in a superheterodyne radio receiver. It appears at the output (O,2) of the frequency changer. The transfer is achieved by a frequency changer circuit (CCT).

(2) A statement used in many computer languages. A reserve word signifying the beginning of a conditional statement.

IFCS International Federation of Computer Sciences. A collection of organizations concerned with the computer sciences, one representing each member country.

IFF Identification Friend or Foe. A system used to indicate whether an aircraft is an ally or an enemy. In operation the friendly aircraft is equipped with a transponder that replies with a coded signal to an interrogation initiated by the IFF system.

IFIPS International Federation of Information Processing Societies. A collection of organizations concerned with information processing, one from each member country.

IFRB International Frequency Registration Board. A permanent body of the International Telecommunication Union (ITU) established in 1947. By maintaining a master list of radio frequencies (RF), it tries to prevent a country using a new frequency that would interfere with existing radio services.

IF-THEN-ELSE A program statement often used in high-level computer languages. If the IF statement is correct, a THEN statement has the meaning that a following expression is to be executed. When the expression is untrue, the ELSE expression is executed.

IGFET Insulated-Gate Field-Effect Transistor. A field-effect transistor that is constructed with a thin dielectric barrier, used to isolate the input electrode (gate) from the channel (CHNL,(h)). *See Figure I.1.* When a control voltage is applied to the gate, it induces an electric field across the dielectric barrier and modulates the nature of the channel region. There are two basic structures of IGFETs, p- and n-channel devices, and there are also two basic classifications of modes of operation, *enhancement* and *depletion*.

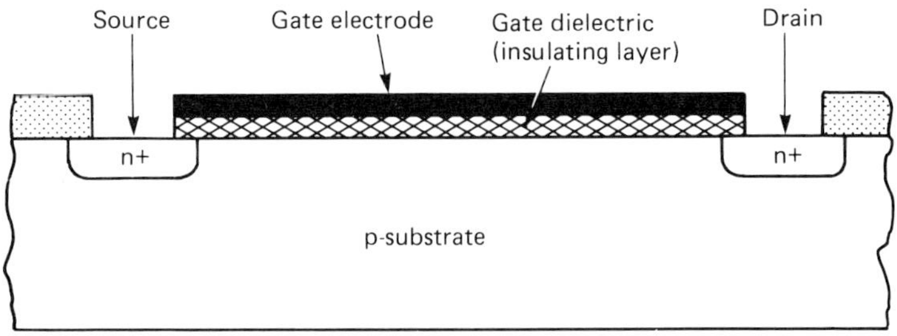

Figure I.1 A cross-sectional view of an insulated-gate field-effect transistor (IGFET)

IIC International Institute of Communications. Providing analysis of issues relating to communication and particularly electronic communication, the IIC is UK-based but has international connections. It aims to provide governments and industries with relevant information.

IIL or **I^2L** Integrated Injection Logic. A family of bipolar integrated circuits (BIC) evolved from diode transistor logic (DTL). They are very compact and provide high-speed operation with a high functional packing density. *See Figure I.2.* I^2Ls use less power, are more stable and cheaper, can be easily mass-produced and are two orders of magnitude faster than transistor-transistor logic (TTL).

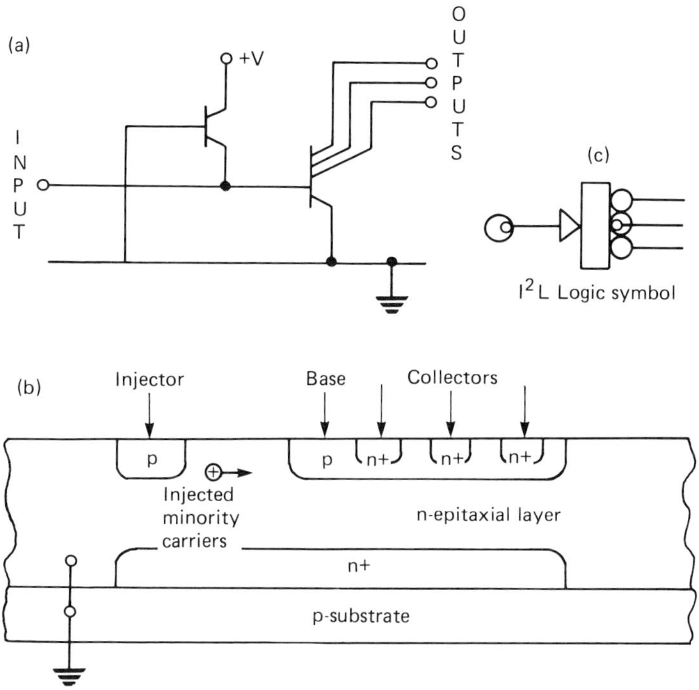

Figure I.2 (a) Equivalent circuit of a single integrated injection logic (I^2L) gate. (b) A cross-sectional view of a basic integrated injection logic-gate structure. (c) I^2L logic symbol

IIR or **I^2R** Rate of energy loss. In an electrical conductor of resistance R carrying a current, I, I^2R is an expression of the rate of energy loss, i.e. the heating effect of a current. It also represents the energy loss.

IL IdLe. A period during which a computer (CMPT) or system stays switched on and available but is not in operation.

I^{2}L *See* IIL.

ILD Injection Laser Diode. A semiconductor diode with a more directional light emission pattern and a narrower spectral width than a light-emitting diode (LED). It is ideally suited to, and commonly used for, optical communications.

ILL Inter-Library Loan. Refers to the information interchange between libraries. The information can be in the form of computer hard copy or computer output microfilm (COM).

ILLIAC ILLinois Institute of Advanced Computation. A faculty of the University of Illinois concerned with the design and implementation of a number of advanced and current technology computers.

ILS Instrument Landing System. Aircraft landing system employing line of sight (LOS,2) frequencies. It incorporates a localizer providing left/right guidance, a guide slope providing up/down guidance and a marker beacon to define progress along the approach course.

IM (1) Item Mark. A special character or item of information used as an indicator to designate the beginning or end of a set of one or more characters (field) that have related data or information. The information concerning a particular object, event, transaction or operation.

(2) Intensity Modulation. In an IM fibre-optic communications (FOC) system, information is transmitted when a modulating electrical signal provides optical sources of variable intensity. De-modulation is achieved with a photo-detector whose electrical output (O,2) is proportional to that of the optical input.

IMD InterModulation Distortion. A particular distortion caused by the non-linearity of the input/output (I/O) characteristic of a component or system, producing new signals at combination frequencies of the two or main sinusoidal input (I/P) signals.

IMP Interface Message Processor. A computer that transmits data by means of addressed packets that occupy the channel (CHNL(b)) just for the time taken to transmit the packet. It is used in an experimental data communication network funded by the Advanced Research Projects Agency (ARPA).

IMPATT IMPact ionization Avalanche Transit Time. A diode (D,5) that acts as a powerful source of microwave (μW) power. The diode is mounted in a microwave cavity and, as part of an appropriate circuit, spontaneous oscillation will occur.

IMS Information Management System. A group of procedures organized as a system that is able to carry out a number of processes upon information, including maintenance and organization cataloguing, locating, storage and retrieval.

INFCO INFormation COmmittee of the International Standards Organisation. One of the committees that advise the International Standards Organisation (ISO).

INFOCOM INFOrmation COMmunications. A message switching service to be used intra-company, supplied in the USA by Western Union.

INFOL INFormation Orientated Language. A high level computer language (HLL) structured around the provision of information. It permits notation to be used by operators with which they are already familiar.

INFRAL INFormation Retrieval Automatic Language. A computer language that permits the user to produce lists of groups of documentation information from indexed data.

INIS International Nuclear Information Service. Offering on-line and off-line searching, INIS uses stores of information held on computer files that relate to aspects of nuclear science and technology.

INMARSAT INternational MARitime SATellite organization. A body or group which administers satellite communications to ships at sea. Vessels are fitted with a special antenna which always points to the satellite despite the movement of the ship.

INSIS INter-institutional integrated Services Information System. A proposed information system to combine a number of services, including electronic mail, facsimile (FAX), telephone, telex (TX), teleconferencing and word processing (WP).

INSPEC INformation Services: Physics, Electrical and electronics, and Computers and control. Available for on-line searching, these stores of information held on computer files contain data relating to physics, electrical, electronic, computer and control subjects.

INT (1) INTerrupt. (a) Normally initiated by an external signal, a discontinuity in the normal flow of a system or routine, that can be resumed at a later time. (b) A specific control signal that diverts the attention of a computer from the main program.

(2) INTeger. An arithmetic function in beginners' all-purpose symbolic instruction code (BASIC), it determines the nearest whole number to x, rounding downwards. For example:

INT (3.3) = 3
INT (12.7) = 12
INT (−8.5) = −9

INTEL INTegrated ELectronics. A major manufacturer of semiconductor devices. Formed in 1969 it produced the first microprocessor (MP).

INTELSAT INternational TELecommunications SATellite consortium. A group of governments and their particular telecommunications authorities formed in 1964 to create a worldwide satellite system. Now over a hundred countries are members. Its first satellite was INTELSAT I known as Early Bird.

INTERMARC INTERnational MAchine Readable Catalogue. A machine readable cataloguing (MARC) system that is made available to a number of countries.

INWG International Network Work Group. An advisory body within the International Federation of Information Processing Societies (IFIPS) that is concerned with research into network protocols and interworking.

IO Integrated Optics. A general term given to monolithic circuits (CCT) that incorporate both optical sources and detectors together with electronic components. Also known as optoelectronic integration.

I/O Input/Output. Relates to the devices, media or operations that are employed to pass information in or out, or both, of a computer or system.

IOC (1) Input/Output Controller. A device that controls interactions between a processing unit and input/output (I/O) equipment. There are two portions to the controller; one controls actions of tape, card and printer equipment and the other is designed for the random processing of records stored on direct access devices.

(2) Integrated Optical Circuit. Consisting of miniature solid state optical components, an IOC is a circuit (CCT) or group of interconnected circuits, generally manufactured on chips (CHIP).

IOCS Input/Output Control System. (a) Computer software specifically designed to control input/output (I/O) operations and communications between peripheral equipment and a central processor unit (CPU). The main areas of communications that are concerned include device allocation, error recovery and file identification. An IOCS comprises a set of computer routines that can also direct other functions such as error correcting, checkpoint label processing and restart. (b) A number of routines which a system analyst or programmer can use in a source program for handling input and output for tapes, cards and printers.

IOP Input/Output Processor. A device that is capable of handling normal data input/output (I/O) control and sequencing.

IOR Input/Output Register. A memory location that accepts information from outside a computer at a particular speed and presents the information to a computer calculating unit at a normally higher speed.

IP Information Provider. In the context of viewdata systems, this is an organization or an individual that provides data for the systems store of information held on computer files.

I/P Input. Relates to devices, media or operations that are employed to pass information into a computer or system.

IPC Industrial Process Control. A data acquisition and control system that can provide great flexibility, as it can accept various types of process data, and a number of output (O,2) signals and format of data that a computer may exercise.

IPL (1) Information Processing Language. A particular computer programming language that provides for the processing of data representing information and the defining of the meaning of the processed information.

(2) Initial Program Load (Loader or Loading). An initiation process that allows a computer program or operating system (OS) into a computer with data records. Once the self-loading routine is established in memory, the program can proceed under its own control, loading and executing other programs.

(3) Initialize Program Load. A routine of initialization used in various intelligent controllers and some computers.

I-POT Inductive POTentiometer. A precision-wound toroidal autotransformer having two sliding contacts, a main contact and an auxiliary contact. During use the auxiliary contact is set to an initial calibration, while the main contact produces an output (O,2) voltage (V,2) that is relative to a mechanical rotary input.

IPS Inches Per Second. A measure of speed, normally used to indicate the rate at which a magnetic tape (MAGTAPE) passes a magnetic head.

IPSS International Packet Switching Service. A public automatic switched data service that provides access between UK data terminals and systems overseas.

I^2R *See* IIR.

IR (1) Infra-Red. Energy that radiates, mainly heat, at wavelengths greater than visible red.

(2) Instruction Register. A device that temporarily stores the current instructions which are being decoded and executed by the central processor unit (CPU) section of a computer.

(3) Information Retrieval. The methods, techniques and processes employed in searching particular computer files for specific items of information, in response to specific user needs.

(4) Current (I), resistance (R) compensation. A technique implemented by the use of operational amplifiers (OP AMP)

forming a closed-loop motor speed control circuit.

IRAS Infra-Red Astronomical Satellite. A satellite in a 900-kilometre-high (560 miles) near-polar orbit, launched in 1983 and circling the Earth 14 times a day. It carries a telescope cooled with liquid helium to perform the first survey of the cosmos at infra-red (IR,1) wavelengths (from 8 to 120 microns).

IRE Institute of Radio Engineers. A professional body in the USA concerned with radio equipment.

IRG Inter-Record Gap. A gap of 0.5–0.75 inch between blocks of information on magnetic tapes (MAGTAPE).

IRIA Institut de Recherche d'Informatique d'Automatique. A French scientific research establishment.

IRL Information Retrieval Language. A computer programming language that is used for the cataloguing of large amounts of data all related to one specific area of interest. It allows for the retrieval of all or part of this data quickly and precisely.

IRS Information Retrieval System. A system which provides particular items of data requested by users. Modern systems employ files of information that can be searched for specific items, in response to the definitions of particular user needs. *See* CAR,2, computer-aided retrieval and DRS, document-retrieval system.

IRW Inverted Rib Waveguide. A variation of the buried heterostructure (BH) semiconductor laser (*see* LASER). Its purpose is to achieve a stable and reproducible laser in operation.

IS (1) Information Separator. A special character used to designate the logical boundaries between items of information in data records.

(2) Indexed Sequential. A method in which computer files are organized on a direct access (DA,2) storage device. A directory is compiled with the address of each record along with its identification, thus permitting the fast access of a relevant record.

ISAM Indexed Sequential Access Method. A method that provides keyed access for rapid and direct retrieval of information from a fast and continuously updated file of information in a computer system.

ISAR Information Storage And Retrieval. (a) The process of storing and retrieving data. *See* IR,3, information retrieval. (b) A computerized system that provides information to users in response to requests.

ISB Independent Side-Band transmission. The transmission of the frequency bands above and below a carrier frequency, each independent of the modulating signal of the other side band.

The system is limited by the difficulty of separating the two signals at the receiver.

ISBD International Standard Bibliographic Description. A convention employed in describing documents that is recognized internationally.

ISBN International Standard Book Number. A number used to aid the location of a particular book within a library or information retrieval system. Every book published is allocated its own unique number, consisting of ten digits (DGTs).

ISDN Integrated Services Digital Network. (a) A planned UK telecommunications network intended to provide a wide, flexible range of facilities for voice, data and image translation. It envisages communications being at a standard speed of 64 000 bits per second (BPS). *See* LCDDS, leased circuit digital data service. (b) An integrated digital network (IDN) used for multiple services such as telephone and data transmissions.

ISL InterSatellite Link. The transmission of messages between communications satellites.

ISM Industrial, Scientific and Medical. A prefix applied to apparatus used in connection with radio interference studies to describe industrial, scientific and medical apparatus that produce radio frequency (RF) energy.

ISO International Standards Organisation. A specialized agency of the United Nations, founded in 1947, and concerned with international standardization across a wide variety of industries and their products.

ISR Information Storage and Retrieval. (a) The process of storing and retrieving data. *See* IR,3, information retrieval. (b) A computerized system that provides information to users in response to requests. The processes employed in the system use the searching of computer files for particular items of information, in response to specific user needs.

ISSN International Standard Serial Number. A number used to aid communication between periodical publishers and libraries, to assist inter-library loan (ILL) and maintenance of computer files. Every title published is given a unique number consisting of eight digits (DGTs) in two groups of four.

ISU Initial Signal Unit. The first of a group of information and check binary digits (BITs) of a multi-unit signal message used in common-channel signalling (CCS).

IT Information Technology. The techniques associated with handling information. They encompass the use of telecommunications systems and computers in handling and processing data, including the application, acquisition, storage and dissemination of pictorial, audio and numerical data.

ITC International Teletraffic Congress. Formed mainly from representatives of telecommunications authorities and industries worldwide, this organization serves as an arena for those involved with teletraffic.

ITF Interactive Terminal Facility. The capability of a computer system to allow a user at a local or remote terminal to elicit a response to any enquiry made of the system. It also allows for a continuous dialogue between the user and the system and provides for immediate diagnostic assistance to any errors made by the user.

ITS (1) Invitation To Send. A computer process which establishes contact with a terminal to allow the terminal the opportunity to send or transmit a message or information if it has any prepared.
(2) Insertion Test Signals. Television (TV(a)) test waveforms that are inserted in field-blanking to enable the picture to be quantitatively tested.

ITT International Telephone and Telegraphs. A multi-national company involved with electronics and telecommunications.

ITU International Telecommunication Union. A United Nations telecommunications agency that has been established to provide standardized communications procedures and practices including the allocation of broadcasting frequencies and radio regulations on a worldwide basis.

ITV Instructional TeleVision. Usually distributed over specially allocated broadcast television (TV(a)) channels (CHNL(g)) in the US, ITV can also be supplied by closed circuit television (CCTV).

IVPO Inside Vapour-Phase Oxidation. A chemical vapour-phase oxidation (CVPO) process that is used for the production of optical fibres (OF). Dopants are burnt with oxygen and gas to form an oxide stream that is deposited onto the inside of a rotating glass tube. It is then sintered to produce a doped layer of a higher refractive index and drawn into a solid fibre.

J

J (1) A symbol used to represent the imaginary component of the sum of a real quantity or number and an imaginary quantity or number.
(2) A J-antenna is a radiating conductor with a length of about half a wavelength of the radiating frequency. It is fed at one end by a matched feeder a quarter of a wave in length.
(3) Joule. In the Système International d'Unités (SI) system of units, the unit of energy or work done. It is defined as the work

done when a force of one newton (N,1) is deplaced one metre (m,2).

(4) Jump. A specific computer (CMPT) instruction that will cause the instruction at the memory location (address) specified in the jump, to be the next executed.

JACK A term applied to both a single-pin plug and its matching socket. The leads from the microphones and earphones of domestic tape recorders and record players are fitted at their other ends with jack plugs which can be inserted into the appropriate jack sockets on the equipment.

JAN Joint Army Navy programme. A programme set up in the USA to standardize on electronic and other military equipment.

JCL Job Control Language. A computer language (designed by IBM) that interprets user instructions and processes them to instruct the central processor unit (CPU) what to do with the user's program.

JDL Job Description Language. Similar to job control language (JCL), JDL is rather more general in that it covers all aspects of defining a task or job to a computer system.

JEDEC Joint Electron Device Engineering Council. A body that helps determine standards on behalf of the USA Standard Institute (USASI).

JFCB Job File Control Block. A set of memory locations containing information about a particular task or job. The information can be accessed by various programs, thus providing control of the operation of the programs.

JFET Junction Field-Effect Transistor. A particular type of field-effect transistor (FET) where the gate forms a combina-

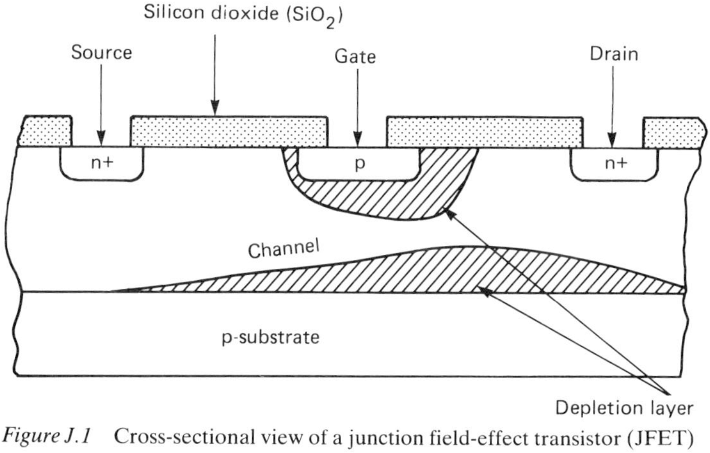

Figure J.1 Cross-sectional view of a junction field-effect transistor (JFET)

tion of p- and n-type semiconductor material (p-n) with the channel (CHNL(h)). A cross-sectional view of a JFET is shown in *Figure J.1*. The channel of the JFET can have more than one gate.

JG Junction Grammar. A model of language employed in analysing sentences, it allows a sentence to be reduced into its syntactic elements. It is used in computerized and other techniques for the automatic translation of text from one spoken language to another; machine-aided translation (MAT).

J-K A special form of flip-flop (FF) circuit that has two inputs, the J and K. *See Figure J.2*. At the application of a clock pulse

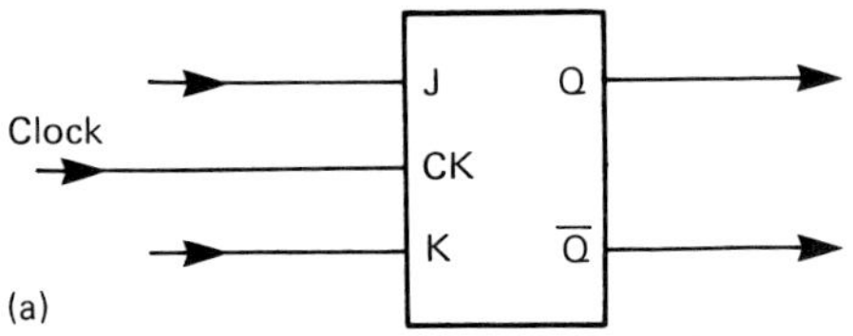

(b)

J	K	Output after one clock pulse	
0	0	No effect	
0	1	Q = 0	$\bar{Q}$ = 1
1	0	Q = 1	$\bar{Q}$ = 0
1	1	Changes Q to opposite state	

Figure J.2 The J-K flip-flop: (a) symbol, (b) truthtable

(CP), a 1 on the J input and a 0 on the K input will set the flip-flop at its 1 state. A 1 on the K input and a 0 on the J input will reset to the 0 state. A 1 simultaneously on both the J and K inputs will cause it to change state regardless of the previous state. A 0 on both the J and K inputs will prevent any change. Also known as a J-K bistable element.

JOHNNIAC JOHN von Neumann's Integrator and Automatic Computer. A computer based on the concepts of John von Neumann. Its particular feature is a program that is stored within the computer and executed under the control of a sequence register.

JOSS Johnniac Open-Shop System. A time-sharing computer language modelled after formula translator (FORTRAN) or beginner's all-purpose symbolic instruction code (BASIC) that has been developed to make quick calculations that are far too complex for a calculator.

JOVIAL Jule's Own Version of IAL. A particular algorithmic language (ALGOL) used by the US Air Force in some of their systems. *See* IAL.

JUGFET JUnction-Gate Field-Effect Transistor. A field-effect transistor (FET) where the gate forms a combination of p- and n-type semiconductor material (p-n) with the channel (CHNL(h)). The channel of the JUGFET can have more than one gate. *See Figure J.1*.

K

k (1) Kilo. A decimal prefix commonly used in association with a base unit in the Système International d'Unités (SI) system of units, indicating a multiplication factor of 10^3 (1000) of that unit. For example, one kilometre (km) equals 10^3 metres (1000 metres).

(2) Cathode. *See* C,10.

K (1) A prefix used in computing to indicate a multiplication factor of 2^{10} (1024); e.g., one kilobyte (KB,3) equals 2^{10} bytes (1024).

(2) A non-standard but often-used symbol for kilohm; used especially on diagrams to replace the decimal point where this could be difficult to distinguish or cause confusion. 2.5 kilohms would be written 2K5, 2K5 ohms or 2K5Ω. The BSI standard representation would be 2.5KΩ.

(3) Kelvin. A temperature scale with a datum point set at absolute zero (where molecular motion ceases) which is 273.15 K below Celsius (C,8) freezing point, 0°. Used for temperature in the Système International d'Unités (SI) system of units. 1 K is an increment similar to 1°C as it has 100 units between the freezing and boiling points of water.

KB (1) KeyBoard. A unit consisting of a group of keys arranged in a particular order (*see* AZERTY, QWERTY, QWERTZ) and used to generate a set of signals for entering data or information. Alphanumeric keyboards may be used for word processing (WP), computer input, text processing and data processing (DP,1); numeric keyboards are used for calculators, touch-tone telephones and accounting machines.

(2) KiloBits. 1024 (2^{10}) binary digits (BITs).

(3) KiloBytes. 1024 (2^{10}) sequences of binary digits (BITs), normally consecutive 8 or 16 bits taken together, operated upon as a unit and usually shorter than a word.

K band K band microwave (μW) frequencies occur between 10.9 and 36 gigahertz (GHz). They form part of a larger continuous range of frequencies covering 0.225–100 gigahertz, which are usually subdivided into bands designated by letters.

kbps KiloBits Per Second. A rate of data transmission of 10^3 binary digits (BITs) per second (1000 bits per second).

KC KiloCycle. A representation previously used to indicate one thousand cycles per second (C/S) now deprecated in favour of kHz.

KCS (1) Kilo Characters per Second. One thousand characters per second is a measurement of data transmission speed that enjoys general acceptance.

(2) Kansas City Standard. The standard for cassette tape recording and playback of data on audio cassettes. KCS is so-called because of a meeting of manufacturers and technical editors in Kansas City who decided on data encoding techniques using standard audio cassettes and recorders.

kg KiloGram. The unit of mass in the Système International d'Unités (SI) system of units. It is equal to the international prototype of the kilogram that is kept at Sevres in France in the form of a piece of platinum-iridium.

kHz KiloHertz. One thousand hertz (Hz).

KIPS Kilo Instructions Per Second. A method of rating the speed of a machine, standing for one thousand operations per second. Large-scale integration (LSI) microprocessors (MP) can execute approximately 500 KIPS, whilst some of the larger computers can execute around 3000 KIPS.

KISS Keep It Simple Sir (or Stupid). Jargon used in certain activities of computer programming.

KIT (1) A colloquial expression used to describe a system that is assembled by the user. It does not necessarily imply an assemblance of components or parts, as it is often applied to a marriage of stand-alone elements, or production equipment.

(2) Key Issue Tracking. A store of information held on computer files in the USA, that provides a continually-updated index of current important topics.

KLIC Key Letter In Context. An index system that lists available computer programs or documents. It is similar to keyword in context (KWIC) indexing system but is based on letters rather than words. Permuted lists of terms are produced, sorted by each letter of every term, with the remainder of the term being entered.

KLU Key and Lamp Units. Equipment that can be installed along with a subscriber's telephone, usually in an office environment, to contend with incoming and outgoing calls on a number of exchange lines, extensions or private circuits.

km KiloMetre. One km is equal to 10^3 metres (m,2) (1000 metres).

KP Start-of-Pulse signal. A term used in the USA to describe a signal that is sent forward down a telephone line to indicate that another signal identifying the party to be called or the onward route of the call is about to be transmitted.

KPIC Key Phrase In Context. An index system that lists available computer programs or documents, similar to keyword in context (KWIC) and keyletter in context (KLIC) indexing systems, but based on phrases rather than words or letters.

KPO Key Punch Operator. A person who controls a device that perforates cards in specific positions. The device works under the guidance of a computer (CMPT) or a keyboard (KB,1) operator.

KSAM Keyed Sequential Access Method. A group of computer library routines together with a structure that allows the storage of data under assigned names. This in turn allows users to obtain records based on the ordering of key field contents.

KSR (1) Keyboard Send/Receive. A combination teletypewriter (TTY) transmitter and receiver that has transmission capability manually from the keyboard (KB,1) only, but no means of automatic origination or retention of data.

(2) Keyboard Select Routing. An application of resource selection where the user requests a connection to a desired resource by the use of a keyboard (KB,1) entry.

KTR Keyboard Type Reperforator. A special kind of teletypewriter (TTY) that receives, punches and types on paper tape (PT). It can simultaneously punch and transmit.

kV KiloVolt. One kV is equal to 10^3 volts (V,1) (1000 volts).

kW KiloWatt. One kW is equal to 10^3 watts (W,2) (1000 watts).

KWAC Key Word And Context. An index system that lists available computer programs or documents. It is similar to keyword in context (KWIC) indexing system, but each significant word is brought to the beginning in alphabetical order. The title has the keyword first, then the words that follow it, followed by the words that came before the keyword.

KWIC Key Word In Context. An index system that lists available computer programs or documents tabulated in alphabetical order of the most informative or significant word in the title of the information. Cf KWOC and *see* KWAC.

KWOC Key Word Out of Context. An index that lists available computer programs or documents tabulated with the title used in full under as many significant or informative words from the title of the information as the compiler considers necessary. Compare KWIC.

L

L (1) Lumens. *See* Lm.

(2) The property of a current-carrying circuit that enables it to generate an electromotive force (EMF) in itself, or a nearby circuit, due to changes in the current (I) and the associated magnetic field.

Lab PS LABoratory Power Supply. A laboratory unit that provides a continuously-variable output (O,2) voltage (V,2) over a specified range. A typical unit includes features such as remote sensing, adjustable over-current and over-voltage protection (OVP), as well as power and overload indicators.

LAMA Local Automatic Message Accounting. A combination of automatic message accounting equipment and automatic number identification equipment housed in the same office. This American system can automatically process a telephone subscriber's dialled toll call for billing purposes without an operator's assistance.

LAN Local Area Network. A computer network operating over a small area linking to computers, electronic mail and word processors (WP) and various other equipment. It can provide communication from site-to-site or provide access to other networks such as the public telephone network and a data transmission network.

LAPADS Lightweight Acoustic Processing And Display System. An antisubmarine system with a microprocessor (MP) unit that converts signals received from sonobuoys into digital (D,4) form. It then carries out all the necessary filtering and analysis as well as processing the data for display (D,6).

LARAM Line-Addressable Random-Access Memory. A block-access memory device consisting of a number of short charge-coupled device (CCD) shift registers that have common input, output (O,2) and regeneration circuits.

LARP Local And Remote Printing. The provision of a computer or word processor (WP) print-out or hard copy near the operator's terminal and if preferred at a more distant point.

LASER Light Amplification by Stimulated Emission of Radiation. An amplifier and generator of a narrow coherent beam of electromagnetic energy in the optical or visible light region of the spectrum. Light from the laser is all the same frequency, has a high energy density and can travel great distances without any divergence from its tight beam.

LASERCOM Light Amplification by Stimulated Emission of Radiation Computer Output Microfilm. A device that combines laser (LASER) and dry film technologies, to enable the laser to mark or write directly onto microfilm. The system results in instant micrographic storage dispensing with any in-between steps.

LB Langmuir-Blodgitt. An elegant technique for producing ordered layers or films or organic molecules. Research is towards the development of chemically- and biologically-sensitive semiconductor devices, integrated optics, luminous displays and solar cells.

L band L-band microwave (μW) frequencies occur between 0.39 and 1.55 gigahertz (GHz) and form part of a larger continuous range of frequencies covering 0.225–100 gigahertz which are usually subdivided into bands designated by letters.

LC (1) Lower Case. A character that can be printed on paper or displayed on a screen, which is not a capital letter and is not included in the upper case (UC) set of characters.
(2) Line Circuit. Equipment at a telephone exchange that acts to supervise the condition of a subscriber's line, when particular control signals are received.

L-C Network containing both inductance and capacitance (C,6). This form of tuned circuit has the product, L-C, of inductance and capacitance constant for a particular frequency.

LCB Line Control Block. An area of a memory containing control information for operations on a communication line. It can be divided into several groups of characters treated as a whole (field) most of which can be designated as generalized control blocks.

LCC Leadless Chip Carrier. A mounting technique employing a high-density package for very-large-scale integration (VLSI) integrated circuits (ICs).

LCCC Leadless Ceramic Chip Carrier. A particular mounting technique employing a high-density package for very-large-scale integration (VLSI) integrated circuits (IC).

LCCD Low Complexity Colour Display. A form of high-quality colour television display providing well-defined alphanumerics and graphics, high-performance displays driven from a logic source, as well as high-quality legibility and sharper colour reproduction.

LCD Liquid Crystal Display. A display that depends on an external light source for visibility and is based on materials that exhibit a regular crystal-like structure even in a liquid state. A normally transparent liquid is sandwiched between two glass plates that have a conductive material on their inner surfaces. When subject to an external electric field the liquid becomes opaque and reflective, *see Figure L.1.* As in other displays, characters are normally built up from segments. As they require minimal power, LCDs are used in battery devices such as calculators, digital (D,4) watches and portable digital instruments.

LCDDS Leased Circuit Digital Data Service. A service provided by British Telecom (BT) of high-speed, high-quality microwave (μW) data transmission links. The communications, between London and 30 other cities, are the first stage of BT's integrated service digital network (ISDN).

LCL LoCaL. A mode of operation in some systems, in which information entered via the keyboard (KB,1) is displayed (D,6)

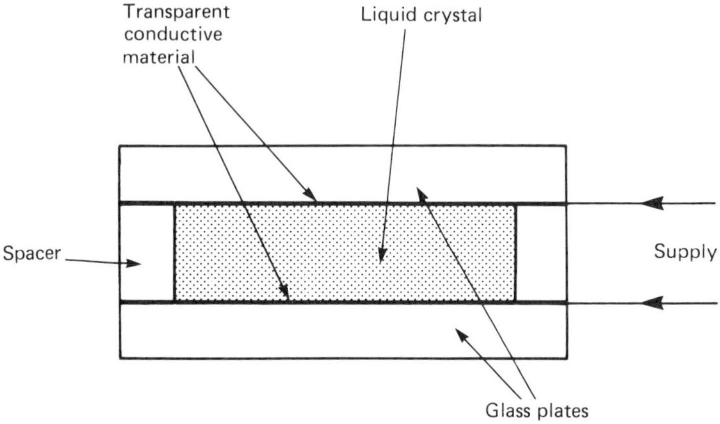

Figure L.1 A simplified cross-sectional view of a liquid crystal display (LCD) cell

on a screen but not transmitted on a communication line.

LD (1) Laser Diode. A junction diode (D,5) that consists of positive and negative charge carrier regions with a junction region which, when injected with electrons, produces radiation at optical frequencies. In materials like gallium arsenide (GaAs) where emitted light is produced rather than heat, some reflected light trapped by polished ends is eventually emitted to produce laser (LASER) action.
(2) Loop Disconnect. A system currently in use on the UK domestic telephone network. Loop disconnect uses a dial tone and up to ten pulses to dial a number.

LDF LoaD Factor. A ratio that is drawn between the number of units of electrical energy during a given period, and the possible number that would have been supplied had maximum demand been maintained.

LDM Limited Distance Modem. An item of data communications equipment (DCE(1)) that has been designed for short-range communication with dedicated lines. Also known as a short-haul modem (*see* MODEM).

LDR Light Dependent Resistor. A semiconductor device with an electrical resistance that decreases with the amount of light falling on it. They have obvious applications in a wide range of devices controlled by light.

LDRI Low Data Rate Input. A comparative term relating to the speed with which a device can receive and take in information.

LDX Long Distance Xerox. A specific form of communication which combines facsimile (FAX) and xerographic copying.

LE (1) Less than/Equal to. A comparison of two values that states whether one value is less than or equal to another. Normally it is used in order for another process or calculation to be carried out. The symbol for LE is $\leqslant$.

(2) Local Exchange. A telephone exchange where subscribers' lines are terminated and connected.

LEA Longitudinally Excited Atmosphere. A type of laser (LASER) that uses gas as an active medium and has an electric field that excites the gas in the direction of its flow, i.e. longitudinally.

LED Light-Emitting Diode. (a) Widely used as indicators in panels and displays, an LED is a p-n junction (p-n) that emits visible light at relatively low power levels when conducting. These diodes (D,5) with a light-producing characteristic provide light of different wavelengths (colours) by variations in their construction. (b) The symbol LED represents a light-emitting diode on parts lists and schematic diagrams.

LEN LENgth. A function in beginners' all-purpose symbolic instruction code (BASIC) computer programming language. It is used to determine the number of characters in a string.

LEO Lyon's Electronic Office. The first commercial computer (CMPT).

LET LETters shift. A function initiated by a letters shift character and performed by a teletypewriter (TTY). It causes the machine to make a physical movement to lower case (LC) so that letters and lower case characters can be printed.

LEXIS (1) LEXicography Information Service. A machine-aided translation (MAT) system in West Germany that provides translation between English, French, German and Russian. It consists of an automatic dictionary with a computer output microfilm (COM) facility.

(2) A UK on-line legal information retrieval system of US origin which contains all Statutes, all Statutory Instruments at any time in force, and all reported cases since 1945.

LF (1) Line Feed. A specific control character used to advance one printing or display (D,6) position to the next position.

(2) Low Frequency. A frequency between 30 kHz and 300 kHz, forming part of a continuous range of frequencies. LFs are internationally agreed to be radio frequencies in band number 5 and have wavelengths between 1 km and 10 km.

LFC Local Form Control. Sited locally, this is a system for off-line data-entry operations by diskette storage of information and unchanging descriptions or specifications (fixed formats).

LFU Least Frequency Used. A replacement process or set of rules leading to a desired output (O,2) from a given input, where existing data in a storage area or memory are replaced by new data, the replaced data being the least frequency used items. When it is required to overwrite information already resident in a main memory with new information, the LFU algorithm is used to decide which page or segment of memory is used least.

LHOTS Long-Haul Optical Transmission Set. Capable of transmitting lightwave signals for long distances via an optical cable, this type of optical link is used for telecommunications between distribution frames or switching centres.

LID Leadless Inverted Device. Also known as a ceramic flipchip. A ceramic block in which is mounted the semiconductor chip (CHIP). Wires from the chip are bonded to the metallized lands. The LID is then positioned over pads on the substrate, and a connection is made with application of heat. *See Figure L.2.*

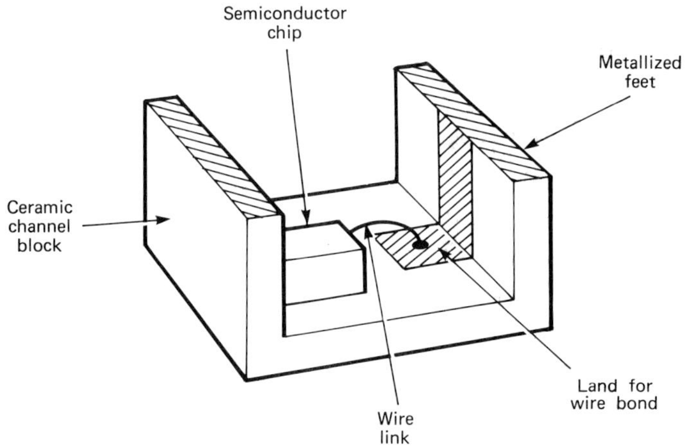

Figure L.2 A leadless inverted device

LIFO Last In First Out. (a) A method of queuing where the last item attached to the queue is the first to be withdrawn. (b) A discipline of queueing in a computer where the newest entry in a file is the first to be removed.

LILO Last In Last Out. (a) A method of queueing where the last item attached to the queue is the last to be withdrawn. (b) A discipline of queueing in a computer where the newest entry in a file is the last to be removed.

LinCMOS LINear Complementary Metal-Oxide Semiconductor. A silicon-gate complementary metal-oxide semiconductor (CMOS) process technology proprietory to Texas Instruments.

It combines the low power, low voltage (V,2) and high input impedance of CMOS integrated circuits (IC) with offsets and voltage swings better than most bipolar devices, and is comparable to some (BIFET) devices.

LIPL Linear Information Programming Language. A computer language that enables the user to employ a notation with which he is already familiar.

LIS Large Interactive Surface. The automated drafting table employed in plotting and/or digitizing drawings in computer-aided design/computer-aided manufacture (CAD/CAM).

LISN Line-Impedance Stabilization Network. A network that simulates the normal impedance of a typical power source impedance. It is used in measuring conducted emissions emanating from equipment to a power line.

LISP LISt Processing. A computer language developed to manipulate symbolic strings of repetitive data and used primarily for research rather than for production programming. It is used generally in the verification of mathematical proofs and for developing higher-level languages.

LLGA Leadless Land Grid Array. A particular mounting technique employing a high-density package for very-large-scale integration (VLSI) integrated circuits (IC).

lm LuMen. Le Système International d'Unités (SI) unit of light flux which is $1/(4\pi)$ of the total light flux that is emitted by a source of 1 candela (CD) intensity.

lm-hr LuMen-HouR. The unit quantity of light that equals one lumen (lm) of luminous flux flowing for one hour.

LMS Level Measuring Set. Instrumentation consisting of a receiving circuit calibrated in decibels (dB), which is used to measure the level of test-signals in telecommunication lines.

lm-sec LuMen-SECond. The unit quantity of light that equals one lumen (lm) of luminous flux flowing for one second (s).

LOC Large Optical Cavity. A laser diode (LD) in which the material junctions are so arranged to provide a wide optical cavity for light waves to reflect back and forth, to provide the laser (LASER) action. The cavity has a higher refractive index than the material surrounding it. The device provides an output (O,2) beam and power greater than a light emitting diode (LED).

LOCAL Load On CALl. If a particular program is too big to be loaded into the core memory of a central processor unit (CPU), sometimes it can be split into subprograms which can be held on disk or magnetic tape (MAGTAPE) which can be loaded into memory on call.

LOCAP LOw-CAPacity. Low-capacity, low-loss paired telecommunication cable able to transmit a pulse-code modulation (PCM) carrier at 6.3 Mbps; used by American companies operating the Bell telephone system.

LOGO A computer programming language. Logo was conceived by the academic Seymour Papert as a computer language for primary school children. Although essentially a dialogue of the list processing (LISP) artificial intelligence language, it exists as a complete language in its own right. What is usually provided under the name LOGO is the excellent picture-drawing facilities known as Turtle Graphics.

LOS (1) Loss Of Signal. The reduction in amplitude of a transmitted signal such that it can no longer be received.
(2) Line Of Sight. A radio or optical link that has an unobstructed straight-line path between its transmitting and receiving antennae. In practice the line is not straight as atmospheric refraction has to be taken into account.

LP (1) Line Printer. A high-speed printer capable of printing an entire line of data simultaneously, 80 to 120 characters long, and then advancing to the next line. These printers normally use rotating drums or chains on which the necessary characters are located. They are normally associated with computers or medium to high-speed terminal services and are much faster than character-at-a-time (*see* CPS) printers.
(2) Linear Programming. A technique employed in mathematics and operational research for determining or solving particular problems involving many variables, where it is required to find the best value or best set of values. The LP technique is not to be confused with computer programming, although problems using the technique can be programmed onto a computer. The mathematical technique used employs a series of linear equations to best allocate any limited resources.

LPC Linear Predictive Coding. A technique used for the analysis of speech and its conversion into digital (D,4) coded information.

LPE Liquid Phase Epitaxy. A semiconductor fabrication process in which a substrate is exposed to a succession of liquids from which thin epitaxial layers of differing compositions are deposited.

LPM Lines Per Minute. (a) Usually used as a suffix forming an expression indicating the number of lines of text that can be dealt with by a printer or a teletypewriter (TTY). (b) Refers to the speed of operation of a line printer (LP).

LRC Longitudinal Redundancy Check. (a) A telecommunications error control based on the formation of a block check that follows preset rules. Information is accumulated at both sending and receiving stations during the transmission of a block. The accumulation is termed the block-check character (BCC) and is transmitted as the last character in the block. A transmitted BCC is compared with an accumulated BCC to determine when a good transmission block has been made. (b) A procedure for checking errors in a computer program.

LRU Least Recently Used. A computer processing routine based on the length of time since the last access to a particular page or segment of main memory was made, in order to determine which part of the memory is to be overwritten.

LS (1) Long Shot. A video or television (TV) term pertaining to the whole of a human being seen on a screen.
(2) LoudSpeaker. An electro-acoustic transducer for generating sound pressure waves from a signal source for radiation over a large area. Most loudspeakers use the moving coil (MC) principle.

LSA Limited Space-charge Accumulation. An LSA diode (D,5) is a transferred electron device (TED) that operates in the limited space-charge mode of a Gunn diode. It can provide the highest peak power of any solid-state device at frequencies up to the millimetre (mm) wavelength region.

LSB Least Significant Bit. The numeral, or binary digit (BIT) in the binary scale of notation, that contributes the smallest quantity to the value of a number and in practice is the right-most digit (DGT).

LSC Least Significant Character. The character in the right-most position in a number or word.

LSD Least Significant Digit. The significant digit (DGT) that contributes the smallest quantity to the value of a number and in practice is the right-most digit.

LSI Large-Scale Integration. The design and production of integrated circuit or microcircuit chips (CHIPs). The solid state circuitry is produced on a single piece of material, normally silicon. LSI denotes circuits of complexity of up to 16 kilobits (KB,2).

LSS Loop Switching System. A term used in the USA to describe a system that accommodates more telephone subscribers than line pairs in the system. It is achieved by concentrating traffic from a remote terminal, onto a smaller number of lines, to a complementary terminal at the local exchange (LE).

LSSD Level Sensitive Scan Device. A sequential device which can be internally initialized via a bit (*see* BIT) stream input.

LSTTL Low-power Schottkey Transistor-Transistor Logic. A widely used and versatile variant of the transistor-transistor logic (TTL) range, with a propagation delay similar to standard TTL (9.5 nanoseconds) but with a much lower power consumption (2 milliwatt).

LT (1) Less Than. A comparison of two values that states whether one value is less than another. Normally it is used in order for another process or calculation to be carried out. The symbol for LT is <.
(2) Low Tension. A low voltage (LV) that is used to supply power for the heaters or filaments of an electron tube. Tubes operated from batteries have an LT of 1.4–2 volts (V,1) and indirectly-heated tubes which usually operate from an alternating current (AC) supply have an LT of 6.3 volts (V,1).
(3) LefT. A command in the Logo (*see* LOGO) computer programming language, to move a Turtle graphics robot through a given number of degrees.

LTRS LeTteRs Shift. (a) A control character that indicates following characters on a teletypewriter (TTY) that are subject to figures shift (FIGS). (b) A physical shift used in a teletypewriter (TTY) that enables alphabetic characters to be printed. (c) The name given to the character that causes the letters shift.

LUF Lowest Useful high Frequency. *See* LUHF.

LUHF Lowest Useful High Frequency. The lowest specific frequency in the high-frequency (HF) band (i.e. 3–30 MHz) that can be used for communication via the ionosphere at any particular time.

LV Low Voltage. Until 1976, LV had stood for voltages (V,2) not exceeding 250 volts (V,1). Now European regulations use the term for voltages up to 1000 V between conductors and 600 V between conductor and earth.

LVA Large Vertical Aperture. A radar antenna that has a large vertical aperture measuring 1.53 metres (m) as opposed to 0.46 m of a linear array antenna. The LVA antenna contains a vertical array of radiating elements that provide more flexibility in the control of the radar's transmit and receive patterns, thereby providing improved performance.

LVD Liquid crystal Visual Display. A low-power-consumption panel that can provide a clear display (D,6) even in bright sunlight. It is a development of a liquid crystal display (LCD) screen that incorporates on a silicon wafer approximately 32 000 dots and a very-light-sensitive large-scale integrated (LSI) circuit.

LVDT Linear Variable Differential Transformer. A device used for measuring small displacements and as such is often employed in the measurement of variables other then position, e.g. weighing.

LVR Linear Video Recorder. A development of video recording which is in principle similar to a conventional tape recorder with stationary heads. Several (incompatible) LVR systems are planned by manufacturers and LVR cassette video cameras have been promised.

LVRT Linear Variable Reluctance Transducer. A close relative of the linear variable differential transducer (LVDT), which consists of two coils and a moveable core, *see Figure L.3*. When

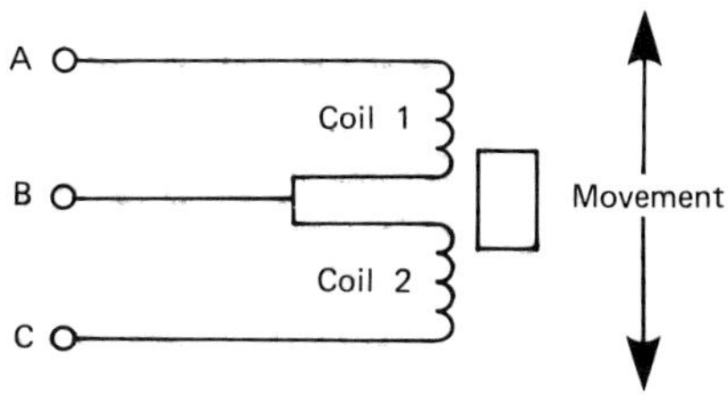

Figure L.3 The LVRT circuit

the core is in equilibrium (null position) the inductance of one coil is increased, and the inductance of the other decreased. An LVRT is normally employed in a bridge circuit connected with a transformer.

LW LongWave. *See* LF(2).

LX LuX. The unit of illuminance in the Système International d'Unités (SI) system of units. It is the quotient of the luminous flux incident on a surface area by that area. It is an absolute unit of illumination equal to one lumen (lm) per square metre (m).

M

μ Micro. A decimal prefix commonly used in association with a base unit of the Système International d'Unités (SI) system of units, indicating a multiplication factor of one millionth, 10^{-6} of that unit. For example, one microsecond (μs) is equal to 0.000 001 second.

μA MicroAmpere. One millionth of an ampere (A), 10^{-6} ampere or 0.000 001 ampere.

μF MicroFarad. An electrical capacitance equal to one millionth (10^{-6}) of a farad.

μm MicroMetre. One millionth of a metre (m), 10^{-6} metre or 0.000 001 metre.

μs MicroSecond. A period of time equal to one millionth of a second, 10^{-6} second or 0.000 001 second.

μW MicroWave. Electromagnetic wavelength shorter than 300 cm (frequencies above 1 GHz).

m (1) Milli. A decimal prefix commonly used in association with a unit of the Système International d'Unités (SI) system of units, indicating a multiplication factor of 10^{-3} of that unit. For example, one millimetre (mm) is equal to 0.001 metre.

(2) Metre (meter, US). The unit of length in the Système International d'Unités (SI) system of units. It is defined as a length equal to 1 650 763.73 wavelengths in a vacuum of the radiation associated with the transition between the levels $2P_{10}$ and $5d_5$ of the Krypton 86 atom.

M Mega. (a) A decimal prefix commonly used in association with a base unit in the Système International d'Unités (SI) system of units, indicating a multiplication factor of 10^6 of that unit. For example, one megavolt (MV) equals 1 000 000 volts (V,1). (b) A prefix used in computing that denotes a multiple of 2^{30}. One Mbyte or megabyte equals 1 048 576 bytes.

MAC (1) Multiple-Access Computer or Multi-Access Computer. A computer system that can receive or despatch input or output (O,2) information from more than one location or terminal.

(2) Machine-Aided Cognition. The application of computers that have artificial intelligence. The system exhibits characteristics commonly connected with human intelligence, including learning, logic, problem solving, reasoning and the ability to understand natural language.

MACRO MACROcode or MACROprogram. A computer source-language statement, routine or program that is able to be understood by a computer compiler or processor with or without the assistance of another routine to decode instructions.

MAGB Microfilm Association of Great Britain. A group that endeavours to increase the standard of the production of microfilm and promote its efficient use.

MAGTAPE MAGnetic TAPE. Available in many forms, magnetic tape normally consists of a plastics substrate on which a magnetic medium is formed from a deposit of an oxide such as ferric oxide (Fe_2O_3) chromium. Recent tapes use particles of metal. The tape may be wound onto reels either in a closed *cassette* container with 4 mm tape or an open *reel-to-reel* with 12 or 25 mm tape. Generally cassettes are used for low to medium quality or domestic audio or microcomputer recording while reel-to-reels are used with high-quality audio or large mainframe computers.

MAI Machine-Aided Index. The use of a store of information held on a computer file to produce an index for documents. The documents are provided with titles that give an indication of their contents.

MAMMAX MAchine-Made and Machine-Aided indeX. A form of machine-aided index (MAI) in which documents in machine readable form are scanned by a computer, and entries are written into the index in a form that complies with instructions in a particular extraction program.

MANIAC Mechanical And Numerical Integrator And Computer. One of the early computers (CMPT). *See* ENIAC.

MAP (1) Microprocessor Application Project. A project launched in 1978 to further the uses of microelectronics for products and processes in the UK. It was initiated and promoted by the Department of Industry and organized by the National Computing Centre.

(2) Memory MAP is one that shows how computer programs are allocated to memory locations and can also indicate where memory exists in the computer.

(3) Bit (binary digit) MAP indicates if a particular item of information is present or absent (is being used or is unused). The item is represented by a single binary digit (BIT), 0 or 1.

(4) Manufacturing Automation Protocol. A worldwide initiative begun by General Motors in the US in order to standardize communications between computers operating in a manufacturing environment. *See also* TOP.

MAP/TOP MAnufacturing office Protocol/Technical Office Protocol. An initiative that enables computers operating in a manufacturing environment to communicate with those operating in technical and office environments. *See* MAP,4 and TOP.

MAR Memory Address Register. A special memory storage area, used to hold the identification of the location of a selected word either to be read from or to be written into the main memory.

MARC MAchine-Readable Cataloguing. Developed by the British Library for the US Library of Congress, this system was designed for the production of records. The records contain information of description of books, articles, reports and documents that are capable of being read by a computer.

MARECS European maritime satellite system. A communication satellite being set up by a provisional organization established by the telecommunications and operating agencies of Europe (Interim Entelsat) to provide telecommunication services for ships at sea.

MARISAT MARItime SATellite service. Making use of geostationary satellites over the Pacific and Atlantic oceans, this service enables communications to be established between ships in either ocean.

MASER Microwave Amplification by Stimulated Emission of

Radiation. A term used to describe a general class of microwave (μW) amplifiers (AMPs) which are based on molecular interaction electro-magnetic radiation. The maser is non-electronic in its working principle, providing low-value noise emissions, especially if operated at very low temperatures.

MASS Multiple Access Sequential Selection. A particular method of storing and retrieving information that is used by a computer.

MAT Machine Aided Translation. A system of translation employing a computer-based multilingual lexicon. The system assists a human translator by providing the swift translation of particular words and/or terms, from one spoken human language to another.

MATV Master Antenna TeleVision. Similar to a community antenna television (CATV), an MATV is designed to provide television pictures but is on a smaller scale than a community system. A typical example would be an antenna serving a block of flats.

MAX MAXimum. The highest or greatest possible size or quantity.

MB MegaByte. A million groups of adjacent binary digits (BITs) operating as a unit (BYTEs). It is employed as a measure of the storage capacity of a device.

MBE Molecular Beam Epitaxy. A semiconductor fabrication process similar to vapour phase epitaxy (VPE). However, this crystal growth method is carried out with the crystalline surface under high vacuum conditions when molecular or atomic beams of the constituent material impinge on the surface of the substrate.

MBM Magnetic Bubble Memory. A memory device employing a very thin layer of magnet garnet set out in its natural state as ribbon-shaped strips or magnetic domains. When an external magnetic field is applied to the material the domains contract into stubby cylinders, which when viewed from the top under a microscope look and behave like bubbles, as shown in *Figure M.1*. When influenced by the external fields the bubbles can be manipulated to represent information binary digits (BITs).

Mbps MegaBits Per Second. A rate of transmitting information equal to 1 000 000 binary digits (BITs) per second.

MC Moving Coil. A type of pick-up used in high fidelity (HI-FI) systems in which the stylus is coupled to a coil assembly which moves between the poles of a fixed magnet.

MCAR Machine Check Analysis and Recording. A diagnostic procedure initiated by an automatic interruption that occurs

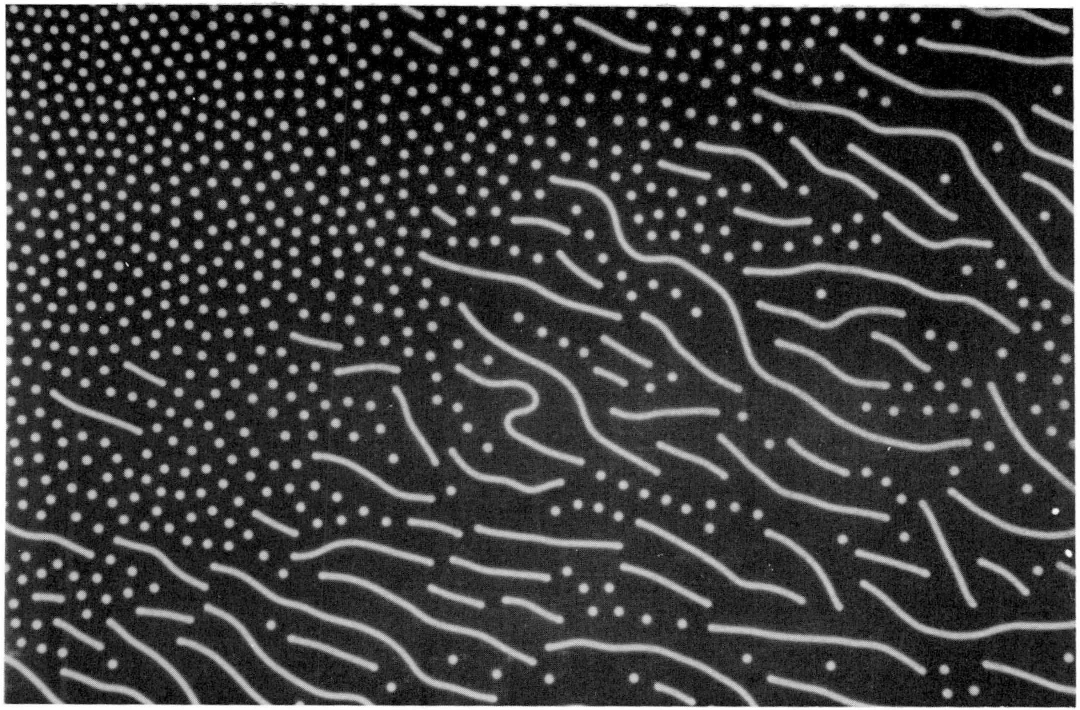

Figure M.1 Bubble and strip domains in a garnet film, employed in a magnetic bubble memory (MBM) (*courtesy* Plessey Microsystems Ltd.)

when the programmed checking of machine functions (machine check) circuits detect a machine error.

MCB Miniature Circuit-Breaker. A device that automatically breaks the normal current flow of a circuit under fault conditions such as a short-circuit, it is rated at around 60 amps (A) or less.

MCH Machine Check Handler. A device that controls the turning on of a protective device when certain conditions arise within a machine. The machine can be programmed to stop, to run a separate correction routine, or to ignore the condition.

MCP (1) Master Control Program. A program that controls all phases of a job. It directs all equipment functions and the flow of all data; provides for automatic error detection and correction; directs the operator with printed instructions and adjusts system operation to changes in system environment.
(2) Message Control Program. A specially designed routine or program used to control the sending or receiving of communications or messages to or from a remote terminal.

MCR Master Control Routine. A supervisory routine in a program consisting of a series of subroutines. It controls the linking of the other subroutines and controls the overlaying of the various segments of memory as required.

MCU (1) Microprogram Control Unit. Used in bipolar microprocessors (MP), this functional unit maintains and generates microprogram addresses. It is also used to control carry-and-shift operations, and in conjunction with an interrupt (INT) control unit sets up the interrupt structure.
(2) Medium Close Up. A television (TV) or video term pertaining to the head and shoulders of a human figure being seen on a screen.

MCVD Modified Chemical Vapour Deposition. A modified form of the inside vapour-phase oxidation (IVPO) process used for the production of optical fibres (OF). Burners travel along the glass tube and soot particles are created inside rather than being deposited by the heat flame on the outer surface, as in the outside vapour phase oxidation (OVPO) process.

MCVF Multi-Channel Voice Frequency. A multiplexing (MUX) method for transmitting a number of signals over one single telephone circuit used by British Telecom (BT).

MCW Modulated Continuous Wave. A particular carrier (CW,1) wave that is modulated by a steady audio frequency (AF) tone.

MD Magneto-Dynamic. A type of pick-up used in high fidelity (HI-FI) systems in which the stylus is coupled to a magnet which moves between the polepieces of fixed coils.

MDA Manufacturing Defects Analyser. A unit that can detect manufacturing defect faults such as open circuits, flow solder and copper etching defects, incorrect component loading as well as component damage.

MDF Main Distribution Frame. A distribution frame used in telecommunications that has connections on either side. The line or lines external to the telephone exchange are connected on one side and the internal cables of the exchange equipment are connected on the other.

MDR Memory Date Register. A special memory storage area that holds the last data word read from or written into the memory location, addressed by the contents of the memory address register (MAR).

MDS (1) Microprocessor Development System. Essentially a system for producing software for microprocessors. A computer-controlled system which employs various devices or peripherals to aid the application development of microprocessors (MP). *Figure M.2* shows what is generally accepted as the configuration of an MDS.

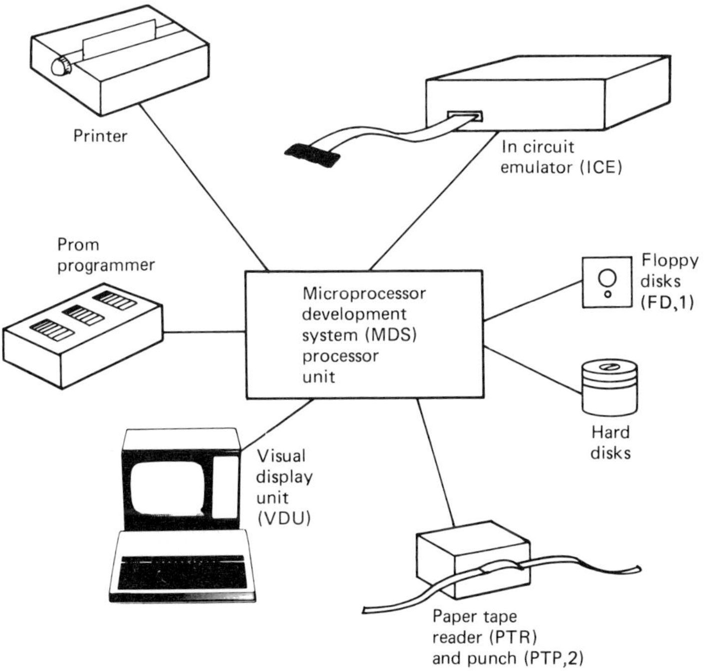

Figure M.2 Typical configuration of a microprocessor development system (MDS)

(2) Minimum Discernible Signal. The least or smallest value of input signal to any device or circuit that is just able to produce a discernible change in the output (O,2).

MDT Mean Down Time. The average period during which a computer is malfunctioning or not operating correctly due to machine failures.

MECL Motorola Emitter Coupled Logic. A family of emitter coupled logic (ECL) devices manufactured by Motorola who, along with Fairchild, were the originators of ECL when they were introduced in the early 1960s.

MEDLARS MEDical Literature Analysis and Retrieval System. A database in the US which contains the names of articles in a number of medical journals published monthly.

MELF Metal ELectrode Face bonding. A particular type of surface mounted device (SMD) package used for cylindrical components. *See Figure M.3.*

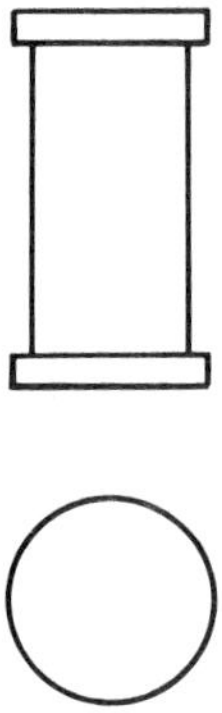

Figure M.3 Detail of a typical metal electrode face bonding (MELF) packaged, surface mounted device (SMD)

MEP Microelectronic Education Programme. A scheme set up by the Department of Industry in the UK to assist, by grant, schools with the purchase of microcomputers.

MESFET MEtal Semiconductor Field-Effect Transistor. A junction field-effect transistor (JFET) that has a barrier formed by a junction between metal and semiconductor (a Schottky barrier) as a gate electrode rather than a semiconductor junction.

METALANGUAGE An artificial computer language employed to define another computer language of which Backus-Naur Form (BNF) is an example.

METAPLAN Methods of Extracting Text Automatically Programming LANguage. A computer programming language used for the retrieval of required portions of text.

METASYMBOL METAlanguage SYMBOL. Any of the symbols employed in the METALANGUAGE artificial computer language.

MF (1) Medium Frequency. Frequencies between 300 and 3000 kHz forming part of a larger continuous range of frequencies. MFs are considered by international agreement to be radio frequencies in band number 6 and have wavelengths of between 100 and 1000 metres (m,2)
(2) Maintenance Free. The absence of the necessity of any activity intended to keep equipment or devices in satisfactory working condition.

MFB Motional FeedBack. A type of loudspeaker system with built-in power amplifiers (AMP,1) and servo-controlled woofer (*see* WOOFER) movements.

MFLOP Million FLOating Point. A measure of computing power that refers to the number of operations per second made when the position of the decimal point of stored numbers changes during computing.

MFLOPS Million FLoating point Operations Per Second. *See* million floating point (*see* MFLOP).

MFSK Multiple-Frequency-Shift Keying. A form of communication where a separate frequency is used to transmit a digital data input code. Ten discrete frequencies are used for the ten numerals of the decimal system of notation. *See also* FSK, frequency shift keying.

MFSS Multi-Frequency Signalling System. A system that transfers information using alternating currents (AC,1) with frequencies in the telephone speech band, to set up, control and release calls. The information is represented by compound signals, made up from a number of voice-frequency currents chosen from a set of available currents.

MFT Multi-programming with a Fixed number of Tasks. The concurrent execution of two or more programs within a single computer, to accomplish a set number of units of work.

MH Manual Hold. When a call is made via the operator in an automatic telephone system, MH is established when the telephone caller's line is held in the busy state until the operator releases it.

MHD MagnetoHydroDynamic. The physical behaviour of electrically conducting fluids (gases or liquids) in the presence of electric or magnetic fields. It can also relate to the observation and study of these phenomena.

MHz MegaHertz. A frequency of 10^6. 1 000 000 cycles per second. *See* mega (M) and hertz (Hz).

MIC (1) Microwave Integrated Circuits. A diverse range of fabrication techniques which range from microstrip circuits through various thin-film technologies (TFT,2) employing glass and alumina substrates, up to monolithic microwave integrated circuits (MMIC) employing bulk silicon bipolar, silicon-on-sapphire (SOS) and gallium arsenide (GaAs) processes. At present the majority of MIC manufacturers produce hybrid devices utilizing a mixture of thin-film and stripline techniques, achieving such circuit functions as oscillators, mixers and amplifiers in combinations to suit many applications and for frequencies up to 60 GHz or more.

(2) MICrophone. *See* MIKE.

MICC Mineral-Insulated Copper-Covered. A type of electrical cable that consists of conductors insulated from one another and the exterior by a mineral material, having an overall sheathing of copper. The cable is often used in harsh or damp environments, employing tight fitting seals (glands) at either end upon entering and leaving enclosures.

MICR (1) Magnetic Ink Character Reader. A device capable of converting into machine code, characters, which are printed in magnetic ink and are readable by humans.

(2) Magnetic Ink Character Recognition. An encoding system, employing magnetic ink, used by banks to enable cheques to be handled automatically.

micro (1) A colloquial expression often used as an abbreviation for a microcomputer or a microprocessor.

(2) Numerical prefix, *see* μ.

MICRO Multiple Indexing and Console Retrieval Operation. Interacting directly with the central processor unit (CPU) of a computer, a MICRO is designed to retrieve, rank and qualify document entries in an index.

MIIS Metal Insulation Insulation Semiconductor. A sandwich of semiconductor materials that employs stored charges between two layers of insulation to produce a nonvolatile memory. When used with the silicon-on-sapphire (SOS) process a memory is produced that is independent of the external power supply.

MIKE MICrophone. A transducer that converts sound energy (acoustic vibration) into electrical energy (signals). It is the initial device used in telephones, in broadcast transmission and all forms of electrical sound recording. Types include carbon, crystal, moving-coil, capacitor (C,6) and ribbon types. *See Figure M.4.*

MIL (1) A unit of length equal to one thousandth of an inch.

(2) MILitary. Usually refers to components, devices or systems produced specifically for or to meet stringent military standards and specifications (SPEC). Normally written as a prefix, for

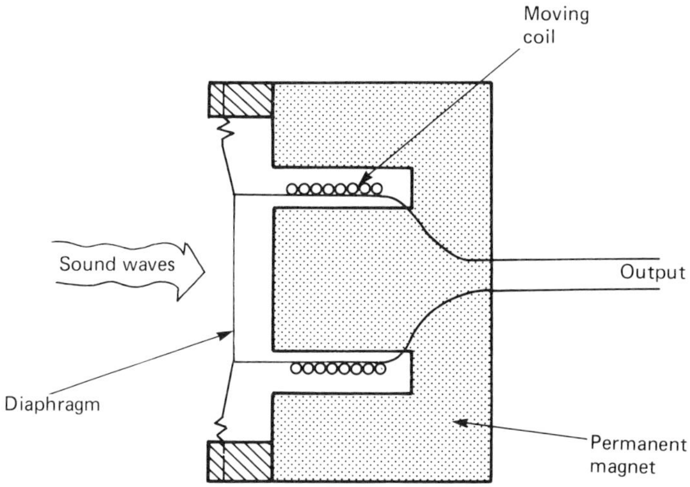

Figure M.4 A cross-sectional view of a moving-coil microphone (MIKE)

example MIL-SPEC for 'military specification' or MIL-STD for 'military standard'. The US Defense Electronic Supply Centre military specification body is responsible for co-ordinating MIL specifications and standards.

MILEPOST Middlesborough Initiative Local Electronic Provision Of Service Technique. Originally an initiative within the area of Middlesborough to take basic National Physical Laboratory Scrapbook Software and add to it additional services in order to produce a more comprehensive computer processing service. It was further developed and launched by British Telecom (BT) in 1983 as a business system. The interactive facilities it offers include information storage and retrieval, word processing (WP), electronic mail (E-COM), multi-activity project management and management information.

MILL Colloquial expression for a central processor unit (CPU) used in some computer systems. A CPU grinds or crunches information as a mill grinds wheat.

milli A prefix denoting one thousandth part of the unit which follows or is attached to it. For example, millisecond is equal to one thousandth of a second. *See* m,1.

MIL-STD-883 MILitary STanDard. The basic US military standard for the reliability of semiconductors. It has three classes: A for aerospace, B for avionics and C for ground use. *See* MIL, 2.

MIMD Multiple Instruction stream, Multiple Data stream. A particular construction or architecture of computers. Multiprocessor systems are MIMDs.

MIN MINimum. The lowest or least possible size or quantity.

MIND Mass-Impregnated Non-Draining compounds. Insulating compounds used in electricity supply cables that control the viscosity of the mineral oil used in the cable and impregnate the insulating papers. Such materials are microcrystalline waxes.

MINIC MINimal Input Cataloguing. A particular system employed for the efficient cataloguing of documents.

MINIDIP MINIature Dual-In-line Package. *See* small outline (SO).

MIPS Millions of Instructions Per Second. A comparative measure of computing power, the number of commands that can be processed per second. 1 MIPS would describe a moderately powerful minicomputer, 20 MIPS would be a medium sized computer and 200 MIPS would be an array processor.

MIS (1) Metal Insulator Semiconductor. A device consisting of a metal top contact layer, a second layer of very thin insulating material and a bottom layer of p-doped (p-type) semiconductor material. An example is a solar cell constructed in a sandwich. (2) Management Information System. (a) A type of data processing (DP) system designed to provide management and other specified supervisory personnel with information of a desired nature. (b) A process of communication where data is processed and recorded and any problems are flagged in order that higher-level decisions can be made, and also so that management can review the overall progress of a project.

MISFEED MISaligned FEED. The failure of a punched card machine (PCM) to feed a card from its hopper; often caused by damaged punched cards.

MISFET Metal Insulator Silicon Field-Effect Transistor. A field-effect transistor (FET) constructed from a wafer of semiconductor material. The wafer has two highly doped regions of opposite electrical polarity diffused into it, forming source and drain regions. An insulating layer is also formed on the surface between these regions and a conductor is deposited on top of this layer to form a gate electrode.

MIST Metal Insulator Silicon field-effect Transistor. *See* MISFET.

MIT Master Instruction Tape. A tape, most often a magnetic tape (MAGTAPE), onto which all the programs for a system of various computer runs are recorded. It is considered to be a fundamental part of an operating system (OS).

MIVPO Modified Inside Vapour Phase Oxidation. A modified form of the inside vapour-phase oxidation (IVPO) process. *See* (MCVD), modified chemical vapour deposition.

MKR MarKeR. Apparatus that controls, selects and sets up telephone traffic paths through a switching stage of a switching centre.

MKS Metre, Kilogram and Second. A system of absolute units where the fundamental units of length, mass and time are the metre (m,2), kilogram (kg), and second (s) respectively. There is also a fourth fundamental quantity for the permeability of free space. The MKS system was used as the basis of the Système International d'Unités (SI) system of units.

MLB MultiLayer Board. A special type of printed circuit board (PCB), with two or more layers of circuit tracks, which is able to increase logic speed and packing density, whilst cutting electrical crosstalk.

MLC Multi-Line Controller. A hardware unit that can be added to an advanced research project agency network (ARPANET) packet-switched computer network to provide the facility to handle up to 64 different terminals at varying signalling speeds (baud rate – *see* BAUD).

mm MilliMetre. One thousandth of a metre (m).

MM Moving Magnet. Another name for a magneto-dynamic (MD) pick-up.

MMF MagnetoMotive Force. The magnetic analogy of electromotive force (EMF). It is the force that gives rise to magnetic fields. Symbol Fm.

MMIC Monolithic Microwave Integrated Circuits. Essentially a microwave integrated circuit (MIC) that employs bulk silicon bipolar, silicon-on-sapphire (SOS), or gallium arsenide (GaAs). First designed in the mid-1970s, it enabled designers to construct architectures that were previously not possible, due to its size and low cost.

MMS Microfiche Management System. A system employed for the storage and retrieval of information held in microfiche. Similar to computer output microfilm (COM).

Mn MaNganese. A grey-pink, hard and brittle metal, atomic number 25, that is often used as an alloying element. It is also used in some forms of primary batteries.

MNOS Metal-Nitride-Oxide Silicon (or Semiconductor). An electrically alterable read-only memory (ROM) in which each individual memory cell consists of a metal-oxide semiconductor (MOS) transistor with an overlaying signal gate electrode, as shown in *Figure M.5*. Some ROMs are preprogrammed by the manufacturer, whereas others are programmed by the user by

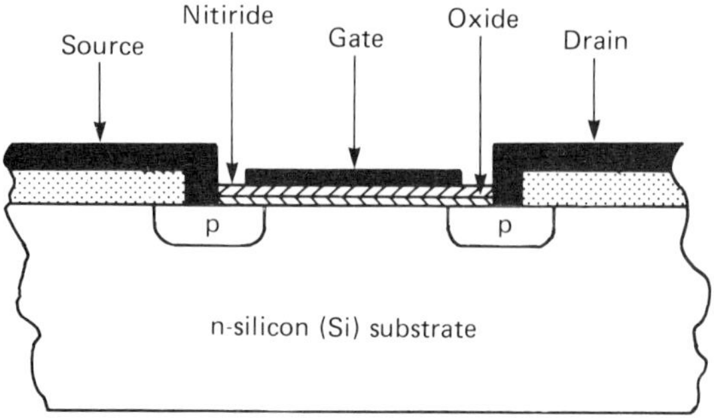

Figure M.5 A cross-sectional view of a unipolar metal-nitride-oxide silicon (MNOS) structure

burning-out selected fusible links (FL) prior to installation. In MNOS structures charge is stored at the interface between silicon nitride and silicon oxide to provide the long-term memory action.

MΩ MEGOHM. An electrical resistance equal to one million (10^6) ohms.

Mo Molybdenum. A metallic element atomic number 42, atomic weight 95.94, melting point 2625°C, boiling point 4800°C and resistivity 4.77 microhm cm^2. It has physical properties similar to those of iron and is used in wire form for the filaments of vacuum tubes, for electrodes of mercury-vapour lamps and for winding electric resistance furnaces.

MOBL Marco Orientated Business Language. A specific computer programming language that permits users to employ a notation with which they are already familiar.

MOCVD Metallo-Organic Chemical Vapour Deposition. A method for depositing very thin and ultra-pure layers of one material onto a substrate. It is well established for the production of gallium arsenide (GaAs) and can be used in the indium phosphide coating process.

MOD (1) MOving Domain. A memory device that is non-volatile like a core memory but much smaller, and is reasonably fast in operation. Unlike a bubble memory it requires no magnetic bias to form and hold the domains and can be batch-fabricated using conventional microcircuit techniques.
(2) MODule. An operating word in the common algorithmic language (COMOL) computer programming language, used to find the remainder of an integer division (*see* DIV,2).

MOD/DEMOD MODulating/DEMODulating. *See* MODEM.

MODEM MOdulator/DEModulator. A unit that modulates and transmits communications signals, and demodulates a signal

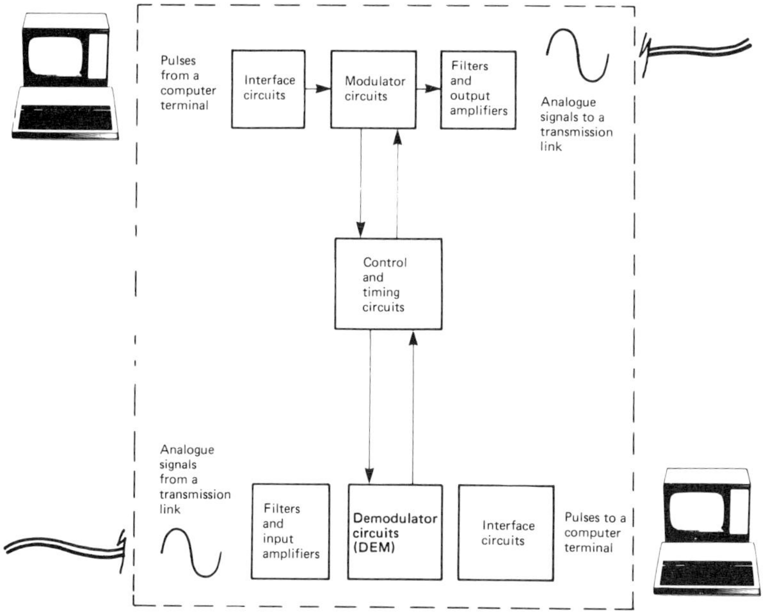

Figure M.6 Typical block configuration of a modulator-demodulator (MODEM)

that is received at a data station or terminal. The term can also be applied to units carrying out other functions like multiplexing (MUX). *See Figure M.6.*

mol MOLe. The unit of amount of substance in the Système International d'Unités (SI) system of units. It is the amount of chemical substance that weighs its own molecular weight in grams (g).

MOL (1) Multiple On-Line programming. A facility providing interactive program development to a number of user terminals simultaneously.
(2) Machine-Orientated Language. A method of notating information that is intelligible to a computer or specific machine. These languages often include information that directly controls machine operations.
(3) Maximum Output Level. The maximum (MAX) signal which can be recorded on a tape without exceeding a permissible distortion level. It is measured in decibels (dB) relative to a reference level.

MOLE ELECTRONICS MOLEcular ELECTRONICS. Various techniques employed in the growing of solid-state crystals to form transistors, diodes (D,5) and resistors (R,2) in one mass for microminiaturization.

MOP Multiple On-line Programming. An interactive set of computer software for the overall control of the operation of a computer system used on International Computer Laboratories (ICL) systems.

MOS Metal-Oxide Semiconductor. The three specific layers of material which are used to form the gate structure of a field-effect transistor (FET). MOS devices tend to have a high capacitance (C,5) n-doped (n-type) silicon substrate which makes them slower than bipolar devices, but they achieve higher functional density (D,3) with fewer process steps which reduces fabrication costs.

MOS DMOS Double-diffused Metal-Oxide Semiconductor. A metal-oxide semiconductor (MOS) device that has had a second diffusion process. This results in a product able to handle higher voltages (V,2) and currents (I) while experiencing very little parasitic capacitance (C,5) and low noise. *See* DMOS.

MOSFET Metal-Oxide Silicon Field-Effect Transistor. A field-effect transistor (FET) constructed from a wafer of semiconductor material with two highly doped regions of opposite electrical polarity diffused into it forming source and drain regions. An insulating layer of silicon oxide is also formed on the surface between these regions and a conductor of aluminium or very highly doped polysilicon is deposited on top of this layer to form a gate electrode. *See Figure M.7.*

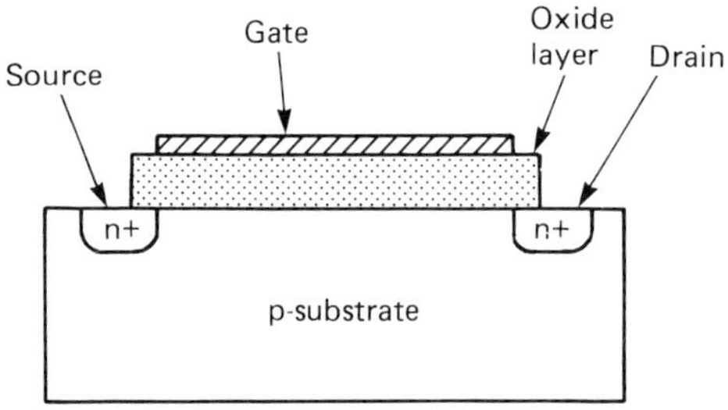

Figure M.7 A simplified cross-sectional view of an n-channel metal-oxide silicon field-effect transistor (MOSFET)

MOSIGT Metal-Oxide Semiconductor Insulated Gate Transistor. Combining polysilicon gate metal-oxide semiconductor (MOS) technology with bipolar power-handling capabilities, the MOSIGT provides around ten times the current (I) rating of metal-oxide silicon field-effect transistors (MOSFET).

MOSRAM Metal-Oxide Semiconductor Random-Access Memory. A random-access memory (RAM) storage medium that employs metal-oxide semiconductor (MOS) transistor cells to store binary 1s and 0s.

MOSROM Metal-Oxide Semiconductor Read-Only Memory. A read-only memory (ROM) storage medium that employs metal-oxide semiconductor (MOS) transistor cells to store binary 1s and 0s.

MOST Metal-Oxide Silicon Transistor. A field-effect transistor (FET) that employs a sandwich of metal and oxide layers. *See* MOSFET.

MP MicroProcessor. Containing all the essential elements of a central processor unit (CPU), this integrated circuit (IC) includes the control logic, instruction decoding and arithmetic processing circuitry. But in order to be useful the microprocessor chip (CHIP) or combination of chips is married with a memory and input/output (I/O) circuit chips to produce a microcomputer.

MPP MolyPermalloy Powder. Powder used to manufacture high flex magnetic powder cores consisting mainly of nickel.

MPS MultiProgramming System. Some computers provide a technique for handling the concurrent execution of a number of routines or programs. The system accomplishes this by overlapping or interleaving their execution, so that one program is executed during the input/output (I/O) waiting time of another program.

MPU MicroProcessor (or MicroProcessing) Unit. The main hardware constituent of a microcomputer. Normally assembled on a printed circuit board (PCB) the MPU is not housed in an enclosure and does not have a power supply. It consists of a microprocessor (MP), a main memory, input/output (I/O) interface devices and a clock (CLK) circuit. There are also a buffer, drive circuits and various passive circuit elements.

MPX MultiPlex. The technique of transmitting two or more messages simultaneously using the same channel (CHNL(a)) or transmission medium. This can be achieved using time division multiplexing (TDM) or frequency division multiplexing (FDM), the two most commonly used techniques.

MPY MultiPlY. An operation performed in computing by two binary digits (BITs) whereby the result is 1 (true), if and only if both digits are 1 (true). If either or both digits are 0 (false), the result is 0 (false). Also known as AND.

MRDF Machine-Readable Data File. A file of information that is capable of being read by a computer.

ms MilliSecond. One thousandth of a second (s), 0.001 of a second or 10^{-3} second.

MS Molecular Stuffing. A clearly defined and ordered process for producing graded refractive index optical fibres (OF). It uses five main steps which are glass melting, phase separation, leaching, dopant introduction and consolidation.

M/S Mark-to-Space. The ratio of the number of times a rectangular wave signal is timed high (on) to low (off).

MSB Most Significant Bit. The numeral or binary digit (BIT) in the binary scale of notation that contributes the largest quantity to the value of a number and in practice is the left-most digit.

MSC (1) Most Significant Character. The character in the left-most position in a number or word.

(2) Main Switching Centre. A fully interconnected switching centre situated in a four-wire switched trunk transit network.

MSD Most Significant Digit. The significant digit (DGT) that contributes the largest quantity to the value of a number and in practice is the left-most digit.

MSDOS MircoSoft Disk Operating System. MS DOS is the version of the personal computer disk operating system (PC DOS) operating system sold by Microsoft to manufacturers, apart from IBM (see IBM). It is operationally similar to PC DOS.

MSI Medium-Scale Integration. The design and production of integrated circuits (ICs) or microcircuits (CHIPs). The solid state circuitry is produced on a single piece of material, normally silicon. MSI denotes circuits of medium complexity, i.e. fewer components in a given area than large scale integration (LSI).

MSS Music Sensor System. A method of directing the cassette transport of a tape recorder on to the start of the next track, or back to the beginning of the one being played. It detects a 40 decibel (dB) drop in signal-to-noise ratio (SNR) between tracks.

MT (1) Magnetic Tape. Available in many forms, magnetic tape normally consists of a plastics substrate on which a magnetic medium is formed from a deposit of an oxide such as ferric oxide (Fe_2O_3) or chromium. Recent tapes use particles of metal. The tape may be wound onto reels either in a closed *cassette* container with 4 mm tape or an open *reel-to-reel* with 12 or 25 mm tape. Generally cassettes are used for low to medium quality or domestic audio or microcomputer recording whereas reel-to-reels are used with high quality audio or large mainframe computers.

(2) Mechanical Translation. The automatic translation from one code, computer language or other system or representation to another code, computer language or system of representation.

(3) Machine Translation. Human language translation performed by computers or similar equipment.

MTBF Mean Time Between Failure. A measure of the reliability of equipment. If the particular unit can be repaired MTBF is the average time between failure; if the equipment cannot be repaired it is the time until the first failure.

MTBM Mean Time Between Maintenance. A measure of the performance of equipment. It provides a guide to how well hardware performs, without necessarily failing completely.

MTF Mean Time to Failure. The average length of time for which a particular system, device, component, or any part or sub-system that can be separately tested, continues to work without failure.

MTL Merged Transistor Logic. A form of logic that employs two multi-collector transistors. It evolved from diode transistor logic (DTL), and is a family of bipolar integrated circuits (BIC) which are very compact and provide high speed operation along with a functional packing density (D,1). MTLs use less power, are more stable and cheaper, can be easily mass produced and are two orders of magnitude faster than transistor transistor logic (TTL). Also known as integrated injection logic (IIL or I^2L).

MTST Magnetic Tape Selectric Typewriter. A typewriter that is able to provide some text handling and editing facilities, but is not as complex as a word processor.

MTTR Mean Time To Repair. A measure of the reliability of equipment or a system, as opposed to mean time between failure (MTBF).

MTU Magnetic Tape Unit. A piece of equipment that records electrical signals on a magnetic tape (MAGTAPE or MT,1), using the tape as a storage medium. The tape, which is flexible, is wound on reels and passes a read/write head. High tape speeds can be used as there is no physical contact between the tape and the read/write head.

MTX Mobile (radio) Telephone eXchange. An exchange that provides an interface between a cellular radio network (CRN) and a public switched telephone network (PSTN).

MUF Maximum Useable Frequency. The highest possible frequency used for radio communication between terrestrial points at one time, via the ionosphere. The MUF is usually the optimum frequency to use for long-distance (DX,3) communications.

MULDEM MULtiplexor-DEMultiplexor (MULtiplexer-DEMultiplexer, US). Analogous to a modulator/demodulator (MODEM), an item of equipment that contains both a digital (D,4) multiplexor (MUX,2) and a digital demultiplexor (DEMUX,2).

MULDEX MULtiplexor-DEmultiplexor. *See* MULDEM.

MULTICS MULTIplexed Information and Computing Service. A multi-user computer system, developed by the Massachusetts Institute of Technology, that is now commercially used.

MUSA (1) Multiple Unit Steerable Aerial. A directive antenna that can be altered or steered to its direction of maximum (MAX) sensitivity by adjusting the phase relationship between the units from which it is assembled.
(2) Multiple Unit Steerable Array. A line of identical rhombic antennas, intended for communications reception, which have their outputs (O,2) combined to provide steerable properties to the array.

MUX (1) MUltipleX. (a) The process of transferring information using a single or common channel (CHNL(a)) or path for the concurrent transmission of more than one information stream. *See* frequency division multiplexing (FDM) and time division multiplexing (TDM). (b) The physical use of one channel (CHNL(a)) or path for the communication of two or more source and sink pairs. *See* space division multiplex (SDM).
(2) MUltipleXor (MUltipleXer, US). A device or technique (often a specialized computer having stored program capability) used to overlap or interleave two or more signals for retransmission at a higher speed. The shared resource for example could be a memory or a bus.

MV Medium Voltage. Now not employed in European standards, MV was a term used to imply a voltage (V,2) between 250 and 650 volts (V,1).

MVT Multi-programming with a Variable number of Tasks. The concurrent execution of two or more programs within a single computer, to accomplish an indeterminable number of units of work.

MW MediumWave. *See* MF(1).

Mx MaXwell. The unit of flux in the now obsolete centimetre, gram, second (CGS) system of units. One maxwell is equal to 10^{-8} weber (Wb).

N

n Nano. A decimal prefix commonly used in association with a unit of the Système International d'Unités (SI) system of units, indicating a multiplication factor of 10^{-9} of that unit. For example one nanosecond (ns) is equal to 0.000 000 001 seconds.

N (1) Newton. The unit of force in the Système International d'Unités (SI) system of units. It is defined as the force required to provide a mass of one kilogram (kg) with an acceleration of one metre (m) per second (s) per second. The unit also relates directly to electrical units as it is the force which, when expended over a distance of one metre, does an amount of work equal to one joule (J).

(2) Negative of sign flag. Usually the most significant bit (MSB), this denotes the sign of a word in two's complement notation.

NA Numerical Aperture. An expression of the light-gathering ability of an optical fibre (OF). It is the sine of the angle of acceptance, θ, which lies between the optical axis of the fibre and the most diverse ray of light that it can accept.

NAB National Association of Broadcasters. An organization that specifies standards, such as the NAB cartridge (a standard endless loop cartridge) or the NAB spool (a 26.6 cm diameter tap spool with a large centre hole).

NAG Numerical Algorithms Group. A body which has provided a collection of library routines for a number of numerical functions for use on a variety of computers.

NAK Negative AcKnowledge. The opposite to acknowledge (ACK). In the method of error control that relies on repeating any message with errors, the return signal that reports the error is NAK.

NAM Network Access Machine. A computer programmed to generate the specific procedures necessary to gain access to a network that connects a number of computers, and to enable the user to interact with the network.

NAMTS Nippon Automatic Mobile Telephone (system). The cellular radio system in Japan.

NAND Not AND. An AND gate that has a negative output (0,2). It is a logic element that has one output and two or more inputs. The output will stand at its defined logic 0 state when and if all the inputs are standing at their defined logic 1 states. *See Figure N.1.*

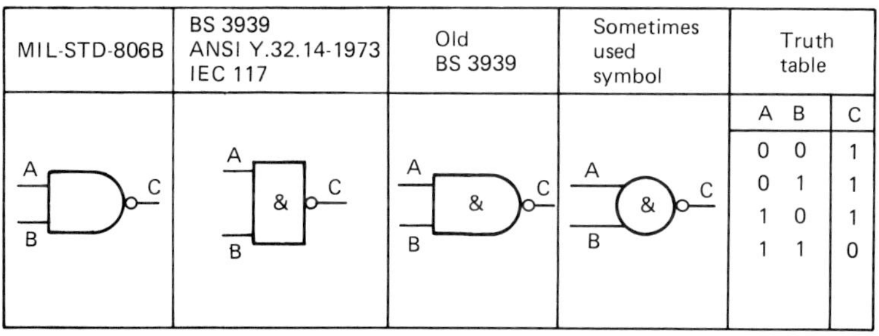

Figure N.1 The NAND gate. In practice the MIL-STD-806B symbol is the one most likely to be met, although the other logic symbols may be encountered

NASA National Aeronautical and Space Administration. One of the bodies that determines standards on behalf of the USA Standards Institute (USASI), covering areas of aeronautics, space science and the related technologies.

NASORD Not in Sequential ORDer. A reference made in computer programming to a file not in sequential order.

NBFM Narrow-Band Frequency Modulation. A frequency modulation (FM) form in which the frequency change is kept at such a level that the signal occupies the same bandwidth as an amplitude modulated (AM) signal carrying the same information.

NBS National Bureau of Standards. One of the bodies that determines standards on behalf of the USA Standards Institute (USASI).

n/c Not Connected. *See* NC(2).

NC (1) Numerical Control. *See* N/C.

(2) No Circuit. A signal used in telephone communications that consists of a low tone (at a frequency of 140 Hz) interrupted at 120 impulses per minute, which indicates to the user that there is no circuit available.

(3) Normally Closed. A switch in which the contacts are closed (contacting) when there is no external force acting upon it.

N/C Numerical Control. The operation of machines by use of punched paper or plastics tape with magnetic spots used to feed digital (D,4) instructions to a machine. The N/C tapes are developed from computer programs.

NCC (1) Network Control Centre. A location in a computer network that controls and in part monitors operations.

(2) Naturally Commutated Cycloconverter. A specialized and restricted AC-to-AC converter having a nonlinear modulating function. It was the first of any type ever to find practical application.

(3) National Computing Conference. Organized by the American Federation of Information Processing Societies, this is an annual conference that takes place in the USA.

NCP Network Control Program. A specific computer program that allows process-to-process communication over a computer network.

NDB Non-Directional Beacons. Specially designed inland beacons used in conjunction with airborne automatic direction finders (ADF). They radiate a signal that exhibits a sharp reduction in strength directly overhead due to the vertical polarization of transmitting and receive antennas, the effect being employed to obtain a fixed position.

NDPS National Data Processing System. A data processing

(DP,1) and transmission service run by British Telecom (BT,1). It provides consultancy to commercial customers in the areas of business information systems, computer output microfilm (COM), information retrieval systems and Prestel.

NDR Non-Destructive Read. A process of reading information in a memory location without erasing or altering that information in any way.

NDRO Non-Destructive Read Out. The sensing of information contained in the internal storage of a computer and the transmission of that information to an external storage unit which inherently does not destroy or erase the record of data that has been read. Reading is usually achieved on tapes, drums, disks and other similar devices, as hard copy on paper or as a display (D,6) on a cathode-ray tube (CRT).

Ne NEon. (a) An inert gas that has the atomic number 10. When ionized it exhibits a characteristic red glow and is used in gas discharge tubes also known as 'glow lamps'. (b) A term used to describe very small glow lamps used in electrical/electronic circuits and devices as indicators. (c) A term used to describe very large gas discharge tubes used for illuminated signs or advertising hoardings.

NE Not Equal to. A comparison of two values that states whether one value is unequal or not equal to another. The expression can be used as a direct comparison but is often used so that another process or calculation can be carried out. The symbol for NE is ≠ or <>.

NEMP Nuclear ElectroMagnetic Pulse. A more correct nomenclature for electromagnetic pulse (EMP), being the very high energy pulse generated by the explosion of a nuclear weapon in the atmosphere.

NEP Noise Equivalent Power. A measurement of the minimum detectable signal that can be found among noise in detection and multiplication processes. It can be defined as the amount of light falling on a photodetector, required to produce an output (O,2) signal equal to the noise output. It is usually expressed in nanowatts.

NEPHIS NEsted PHrase-Indexing System. An automated system of indexing subjects listing all the words in the title of a particular document. It is arranged so that every word appears in turn.

NEPR Nato Electronic Parts Recommendations. International North Atlantic Treaty Organization (NATO) specifications are produced as Nato electronics parts recommendations.

NEQ Not EQual to. *See* NE.

NERFET NEgative Resistance Field-Effect Transistor. This device is a real space transfer 'Gunn oscillator', in which negative resistance occurs in a field-effect transistor (FET) channel (CHNL(h)) as carriers transfer from the two-dimensional electron gas (2 DEG) over a heterojunction barrier.

NEV Non-Equivalent Gate. A gate that has only two inputs and checks for these being the same (i.e. both 1 or both 0). There is no simple relay equivalent to a NEV gate.

(a)

A	B	C
0	0	0
0	1	1
1	0	1
1	1	0

(b)

Figure N.2 (a) The NEV gate logic symbol. (b) The NEV gate truth table

NEXIS Newspaper database, Mead Data Control. (US.) A store of information on computer files providing the full text of newspapers and other agency news. It can be searched directly from a terminal.

NEXT Near-End CROSSTalk. An unwanted energy transfer from a disturbing circuit to a disturbed circuit, where the energy travels in a direction different from that of the signal in the disturbing circuit.

NFB Negative FeedBack. A signal that is returned to an earlier stage of an amplifier (AMP) opposing the input signal at that point. NFB reduces gain in the return or feedback loop, improves linearity and provides more level frequency response. The feedback circuit is built from passive components which are independent of active devices in the feedback loop. It is also known as 'degenerative feedback'.

Ni NIckel. A metal that has the atomic number 28. It is widely used in electronics, in magnetic alloys due to its strong ferromagnetism, and in electrolytic cells as a conductor. *See* Ni-Cad.

NI NonInhibit. Noninhibit interrupts are provided in some computer systems and they include powerfail, auto restart, TTY break, Memory Parity and Protect, Interrupt Program Time Out, and nonimplement Op code.

NIBBLE A colloquial term to describe a 4-binary digit (BIT) word in contrast to a byte, which is generally considered to be an 8-binary digit (BIT) word.

Ni-Cad NIckel-CADmium cell. A secondary cell having potassium hydroxide as an electrolyte, a positive plate (anode) of nickel-oxide/nickel-hydrate mixture, and a negative plate (cathode) of cadmium. It can provide a high rate of discharge along with a steadily maintained voltage (V), properties brought about by the use of cadmium. The cadmium gives the cell a lower charging voltage than a nickel-iron cell (Ni-Fe) and reduces internal resistance.

NICHROME NIckel-CHROMium alloy. A trade name given to an alloy of approximately 63% nickel, 14% chromium and 23% iron. It is used widely for the manufacture of thin-film resistors (R,2), wire-wound resistors, and heating elements, due to its ability to operate at very high temperatures and its high resistivity.

Ni-Fe Nickel-FErrous (iron) cell. A secondary cell capable of delivering heavy currents (I) using potassium hydroxide as its electrolyte. The cell has a positive plate (A,3) made from a nickel-oxide/nickel-hydrate mixture and the negative plate (C,10) is of iron oxide. The cell is lighter, more durable and can operate at lower temperatures (−30°C) than a lead-acid cell, but is more expensive and has an electromotive force (EMF) of only 1.2/1.3 volts (V,1).

NIP Nucleus Initialization Program. Part of an operating system (OS) that resides in main memory or storage, which initializes the resident control program.

NIS NATO Identification System. A system embracing the whole military scene to provide NATO's identification friend or foe (IFF). It provides fully interoperable identification facilities between all NATO forces on land and sea and in the air, as shown in *Figure N.3*.

NL New Line character. A functional character that controls or demands action of a printer or display (D,6), by causing the printing or display position to be moved to the next printing or display line. It combines the functions of a carriage return (CR) and a line feed (LF).

NLQ Near Letter Quality. Dot matrix printers produce characters by forming dots closely together. In general the closer together the dots the better the reproduction (the more it is like typewritten characters). Earlier dot matrix printers had to run slowly for good letter quality, but now high-speed printers are being produced with higher quality characters. They are termed NLQ machines.

nm NanoMetre. One thousandth of a millionth of a metre (m), that is 0.000 000 001 or 10^{-9} metres. It is a convenient unit for designating the wavelength of light. It is also equal to 10 angstroms (Å) and is synonymous with millimicron.

NMC Network Measurement Centre. A location on an advanced

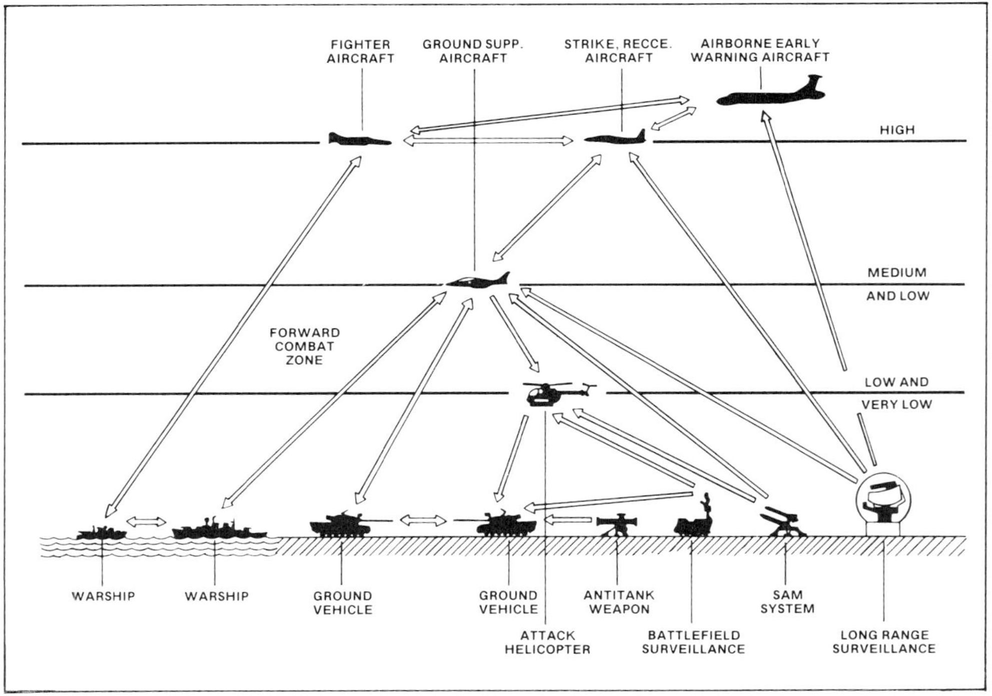

Figure N.3 The NATO identification system (NIS) overall scene (*courtesy* Cossor Electronics Ltd.)

research projects agency network (ARPANET) packet-switched computer network used to measure performance, normally by means of relatively long-term test.

NMI Non-Maskable Interrupt. A signal sent from an input/output (I/O) device or chip (CHIP) to a microprocessor unit (MPU) to request a service (an interrupt – INT) at the highest priority that is not affected by an interrupt mask. It is typically used in the event of a power failure.

NMOS N-type or channel Metal-Oxide Semiconductor. Faster in operation than p-channel metal-oxide semiconductor (PMOS) devices NMOS are the newer variety of large scale integration (LSI) chips (CHIP). They are transistor-transistor logic (TTL) compatible and use one power supply and have gates about half the size of PMOS gates.

NMOS RAM N-channel Metal-Oxide Semiconductor static Random-Access Memory. *See* NMOS, RAM and SRAM.

NMT Nordic Mobile Telephone (system). The cellular radio used in the four Nordic countries.

NO Normally Open. A switch in which the contacts are open (separated) when no external forces act upon the switch.

N/O Normally Open. *See* NO.

NOOP NO-OPeration. *See* NOP.

NOP No-OPeration. An instruction that is used to force a delay of one instruction cycle without altering in any way the status flags or the contents of a memory location.

NOR Not OR. An OR gate that has a negative output (O,2). It is a logic element that has one output and two or more inputs. The output will stand at its defined logic 0 state if one or more of the inputs stand at their defined logic 1 states. *See Figure N.4.*

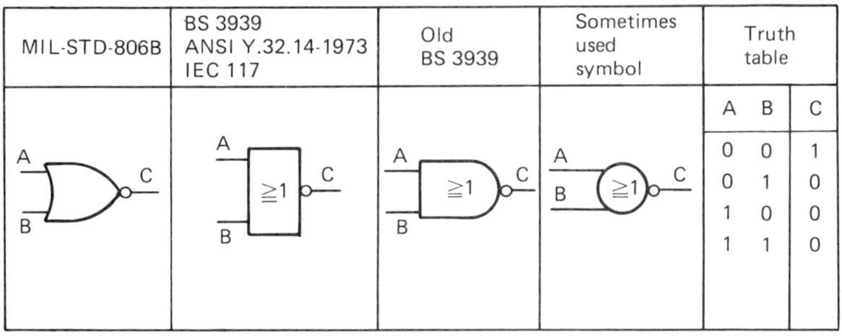

Figure N.4 The NOR gate. In practice the MIL-STD-806B symbol is the one most likely to be met, although the other logic symbols may be encountered

NOS Network Operating System. A generic term employed to describe distributed software for the control of the operation of a computer network.

NOT A logical operator having the property that if A is a statement, then the NOT of A is true if A is false and it is false if A is true.

NOVRAM NOn-Volatile Random-Access Memory. A particular combination of random-access memory (RAM) and read-only memory (ROM) which produces a device with RAM attributes that is unaffected by the loss of power, in that it does not lose its contents. It is therefore termed non-volatile. (*See* NVM, non-volatile memory.)

np or **n-p** N-doped P-doped junction. A region in a semiconductor where the n-doped and p-doped areas meet. It has a high resistance in one direction and a low resistance in the other.

Np Neper. A dimensionless unit used in telecommunications. It is the natural logarithm of the square root of the power ratio and is named after the inventor, John Napier.

NPA Numbering Plan Area. Part of a national or integrated telephone numbering plan, it is an area that consists of further numbered subdivisions that are identified by a particular number-plan trunk code. In the UK, London has been given the digit 1.

npin and **n-p-i-n** N-doped P-doped Intrinsic N-doped transistor. A transistor that is constructed with an intrinsic (undoped) layer of material between its base and collector to extend the high frequency range.

npip and **n-p-i-p** N-doped P-doped Intrinsic P-doped transistor. A transistor that has an intrinsic (undoped) layer between two p-regions (p-type).

npn or **n-p-n** N-doped P-doped N-doped transistor. A transistor that is constructed from two n-doped (n-type) regions with a p-doped (p-type) region separating. The voltage (V,2) polarities and current (I) for an npn are reversed from those of the pnp (*see* pnp), since the conductivity types are interchanged.

NRCd National Reprographic Centre for Documentation. A UK organization devoted to the study of reprographics, publishing information and evaluating equipment.

NRMM National Register of Microfilm Masters. A register of original negatives of microfilms or microfiches housed in the US Library of Congress.

NRZ Non-Return to Zero. (a) A signal presented in the digital (D,4) form having pulses which are full-symbol-length (*see* diagram) so that there is no alteration of signal level when a symbol is repeated, i.e. there is no return to zero level, as shown in *Figure N.5*. (b) A method of writing information on a magnetic surface where the current passing through the write head winding does not return to zero after the write pulse. The

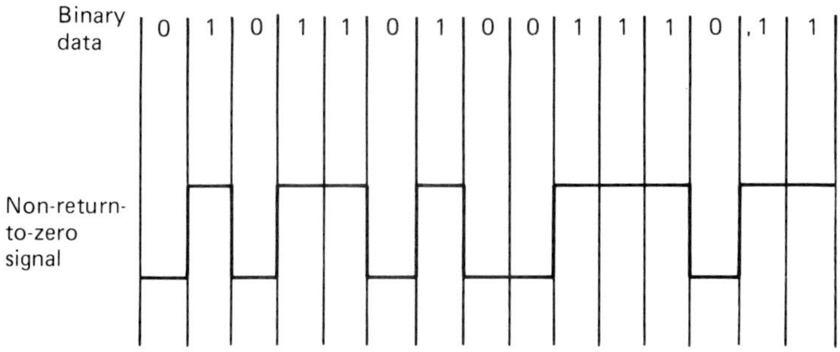

Figure N.5 A non-return-to-zero (NRZ) signal

reference condition is represented by the change in state of magnetization between the 1 and 0 condition.

NRZI Non-Return to Zero Inverted. A method of writing information on a magnetic surface in which the state of magnetization representing either 0 or 1 changes only when a 1 is transmitted.

NRZ(M) Non-Return to Zero (Mark). A method of writing information on a magnetic surface in which a change in the condition of magnetization represents a 1 and no change represents a 0.

ns Nanosecond. *See* nsec.

nsec Nanosecond. One thousandth of a millionth of a second (s), that is 0.000 000 001 or 10^{-9} seconds. It is synonymous with millimicrosecond.

NSI National Supervisory Inspectorate. A controlling authority, with inspectors at local level, which is the supervisory body for the approval and quality assessment of connectors and other components to British Standard 9000 series specifications (*see* BS 9000).

NSW National Software Works. A co-operative venture amongst some associations in the USA, to design a distributed operating system (OS) for a heterogeneous computer network.

NTC Negative Temperature Coefficient. The attribute of a semiconductor material which has a resistance that decreases with an increase in temperature. NTC semiconductor materials are normally employed in thermally sensitive resistors (THERMISTOR).

NTIA National Telecommunications and Information Agency. An agency of the Department of Commerce in the USA.

NTIS National Technical Information Service. Covering technical and scientific reports from US Government agencies, this service generates a multi-disciplinary store of information on computer files.

NTSC National Television System Committee. An American body that specified and pioneered the NTSC compatible colour television system in 1953, the system being used in the USA and in a number of other countries. The Phase Alternation Line (PAL) system in the UK and other countries is an improved development of the NTSC system. The body also determines standards on behalf of the USA Standards Institute (USASI). Also known as National Television Standards Committee.

NTU Network Terminating Unit. Part of telecommunication network equipment that connects directly to data terminal equipment. It operates between local transmission lines and a telephone subscriber's interface.

n-type (1) An extrinsic semiconductor containing a higher density of conduction electrons than of mobile holes (Valence band electrons), which means the electrons are the majority carriers. A semiconductor region is doped to provide the excess of electrons (*see Figure N.6*). *Compare p*-type semiconductor. (2) An early type of microwave connector. It is screw-coupled and has an air interface, the centre conductor being supported by a polytetrafluoroethylene (PTFE) or polystyrene bead.

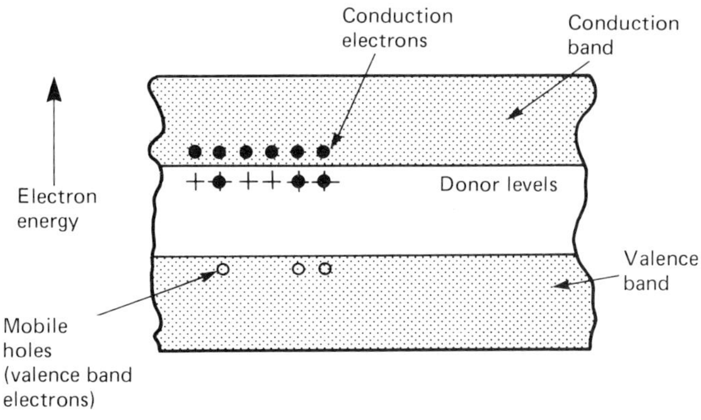

Figure N.6 Energy bands of an n-type semiconductor

nu The nu value or abbé constant is a mathematical expression applied to optical systems for determining the correction required to counteract image imperfections in the system.

NU Number Unobtainable. A tone used in telephone communications indicating to a caller, by use of an audible signal, that a call cannot be completed. There are four possible reasons: the line required is either faulty or out of commission; a non-existent or invalid number has been dialled; equipment has been misoperated or has malfunctioned; access to the required number is banned. The tone is also relayed if a caller does not send an effective signal during a certain period of time.

NUA Network User Address. A set of packet assembler-disassembler (PAD,2) parameters that identifies the computer with which a local area network (LAN) or wide area network (WAN) user wishes to communicate.

NUI Network User Identity. A set of packet assembler/disassembler (PAD,2) parameters that provide a local area network (LAN) or wide area network (WAN) user with a recognizable identity.

NUL NULl (a) The absence of information, as opposed to a zero or blank which indicate the presence of no information. (b) Zero. (c) The deflection from a centre or end position. (d) A computer instruction to do nothing, except proceed to the next sequential operation. Null characters can be inserted into, or removed from, a sequence of characters without the meaning of the sequence being affected although the control of equipment or the format may be altered.

NVM Non-Volatile Memory. A permanent memory storage medium that is able to retain information in the absence of power and allows the data to be available when power is restored. A magnetic memory is an example of a non-volatile memory.

NVROM Non-Volatile Random-Access Memory. *See* NOVRAM.

O

O (1) Oxygen. A chemically active, gaseous element forming one fifth of the Earth's atmosphere. It is odourless, has an atomic number 8, atomic weight 15.9994, melting point −218.8°C and a boiling point −182.970°C.

(2) Output. (a) Results issuing from a computer or device which could be answers to statistical or mathematical problems. (b) Signals produced by a device that are used to drive peripheral devices. (c) Information transferred from the internal storage of a computer to secondary or external storage. (d) A channel for expressing a state on a device or logic element. *See also* O/P.

(3) O network. A network consisting of four impedances connected to form a closed loop or square (*see Figure O.1*). It has an input applied across one element and the output (O,2) taken across the opposite element.

OB Outside Broadcast. Mobile facilities for video or television (TV(a)) transmission or recording.

OBE On-Board Electronics. Demands for higher levels of

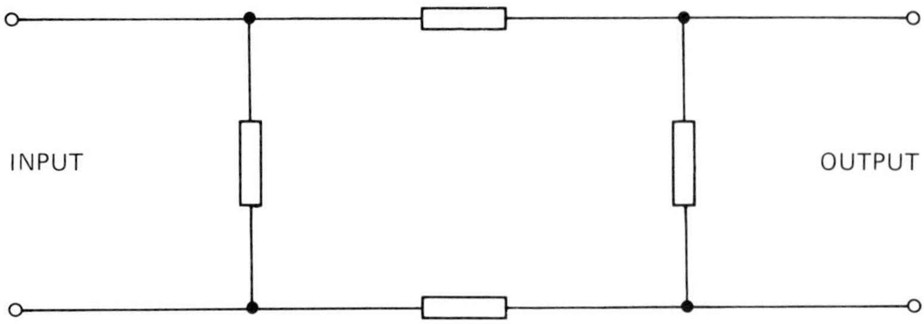

Figure O.1 An O network

component integration have caused light-emitting diode (LED) manufacturers to provide products with drive electronics and LED displays in a single package. Called 'on-board electronics', these units are mostly end-stackable for multicharacter assemblies.

OBIC Optical Beam Induced Current. A method of contactless testing employed in automatic test equipment (ATE), in which a localized current (I) is induced in a semiconductor device in order to examine the internal logic.

o/c Open Circuit. A circuit which is broken and therefore cannot carry a current (I) flow.

OC Optical Cavity. The active area or cavity in a laser (LASER), formed by two heterojunctions (HJ). *See* LOC, large optical cavity.

OCB Oil Circuit-Breaker. A high-voltage (HV) circuit-breaker that has an arc through oil, which raises the temperature of the oil causing bubbles of gas (mainly hydrogen) to be formed. When the current is returned to zero the turbulence caused by the gas and the cooling effect of the surrounding oil disperses and de-ionizes the path of the arc and restores its dielectric strength.

OCLC Ohio College Library Centre. An on-line computer network, with more than 3000 remote terminals, for cataloguing library information. It is available mainly in North America.

OCR (1) Optical Character Reader. A device used to produce coded signals, in digital (D,4) form, from printed characters, numbers or symbols on paper, that are suitable for a given computer. It usually requires source material in a specially designed typeface for maximum efficiency, although OCR recognition abilities are constantly improving.

(2) Optical Character Recognition. Printed or written characters recognized by machines using inputs from optical, photoelectric transducers or other photosensitive devices.

OCR WAND Optical Character Reader WAND. A hand-held optical character reader (OCR,1). An example is a bar code scanner, which is a unit that can read documents encoded in a special bar code, at speeds of hundreds of characters per second (CPS).

OCT OCTal. Pertaining to eight, but usually used to describe a numbering system with the base eight. Numbers in the system express values with each digit position representing successive powers of eight. For instance, decimal 85 would be written as 125, representing 1 times 64, 2 times 16 and 5 × 1. If used as a shorthand method of writing binary digit (BIT) strings, the bit string 001,010,101 can be written as octal 85, whereas in straight binary it is 1010101.

ODB Output to Display Buffer. Peripheral to a computer, this data storage area holds information while it is being transmitted from an output (O,2) to a display (D,6) device.

Oe OErsted. In the now obsolete centimetre, gram, second (CGS) system of units, this is the unit of magnetic field. One oersted is equal to 79.58 amperes (A) per minute.

OE Output Enable. An output (O,2) control signal that provides a completely separate means of controlling the output buffer of a memory device, so eliminating the need for bus connections.

OEM Original Equipment Manufacturer. A manufacturer who uses computers or components as a section or part of the items that he directly sells, as opposed to end users who buy the products for their own use.

OF Optical Fibre. A dielectric waveguide for electromagnetic energy at optical wavelengths. The fibres provide a path for a single beam of light or in multiples, such as the transportation of a complete image. The fibres are provided as either a single fibre or a cable bundle. They may be bent or curved to meet specific needs. *See* FO (fibre optics).

OFHC Oxygen-Free High-conductivity Copper. Pure copper that has 100% conductivity. It is used for the construction of high-power electron tubes as it does not release any appreciable gas when heated.

OF LIN PS Open Frame LINear Power Supply. A low cost, very simple construction power supply unit (PSU) that employs linear techniques. Normally mounted on a single open chassis without a cover, this type of unit has a power rating of up to 150 watts (W,2).

OFPS Open Frame Power Supply. A power supply unit that is normally mounted on a single chassis without a cover. It is a very cost-effective design intended for original equipment manufacturers (OEM) to build into their own equipment.

OFR Oil resistant, Flame Retardant. Refers to a standard of cable suitable for use in situations involving contact with oil and grease and for power and lighting services, reeling and trailing applications or for exterior equipment such as mobile cranes and conveyours.

OF SMPS Open Frame Switch Mode Power Supply. A switch mode power supply (SMPS) that is normally mounted on a single open chassis without a cover. It is a very cost-effective design intended for original equipment manufacturers (OEM) to build into their own equipment. *See Figure O.2.*

Figure O.2 An open-frame switch mode power supply (OF SMPS) (*courtesy* Coutant Electronics Ltd)

OFTF Optical Fibre Transfer Function. A constant for a particular optical fibre (OF). When a light wave enters an optical fibre the fibre brings about a transformation upon it. If the input composition and the transfer function are known, then the output can be determined.

OGM OutGoing Message (tape). Refers to the tape in a telephone answering machine on which the user leaves a message for callers.

OIS Office Information System. A general term that can be applied to any of the electronic systems employed in a variety of office functions. The processes concerned can include information retrieval, telecommunications or word processing (WP).

OL On-Line. (1) When a piece of equipment of a subsystem is connected and powered to deliver its proper output the system is said to be 'on-line'.

(2) Refers to a particular class of equipment capable of interacting with a computer.

OLRT On-Line Real-Time operation. A method of operation whereby results from a computer are obtained during the progress of an event. Input data are passed directly from a measuring device to the computer.

OLTS On-Line Test System. The running of a whole system against test data with a verified solution, or the complete simulation of an actual running system to test out the adequacy of the system. The system runs at speeds compatible with a computer and is connected directly to the main processing unit (MPU).

OMNIX Onyx Microcomputer uNIX. A version of the general-purpose multi-user interactive operating system (OS) (*see* UNIX) available on Onyx microcomputers.

OMR (1) Optical-Mark Recognition. A technique in which a device scans and recognizes graphite marks on a document and converts the information into digital (D,4) signals for input to a computer.

(2) Optical-Mark Reader. A device that recognizes graphite marks, which are normally produced by a pencil on card or paper, and converts the information into digital (D,4) signals for input to a computer.

ONI Operator Number Identification. Equipment at a local dial central office that permits an operator to break-in for sufficient time to identify and acquire the calling telephone number. This is done so that the number may then be dialled into centralized automatic message accounting equipment.

ON-TAP ON-line Training And Practice. A specific set of information held on computer files containing data from a number of other stores of information held on computer file. It is designed to help train personnel in the uses of direct searching of computer files.

O/P OutPut. Relates to devices, media or operations that are employed to pass information out from a computer or system. *See also* O,2.

OP AMP OPerational AMPlifier. Normally an integrated amplifier (AMP) of high gain and wide bandwidth that can be used to perform mathematical operations. *See Figure O.3.*

OP REGISTER OPeration REGISTER. A memory storage location used to hold the operation code of computer instructions.

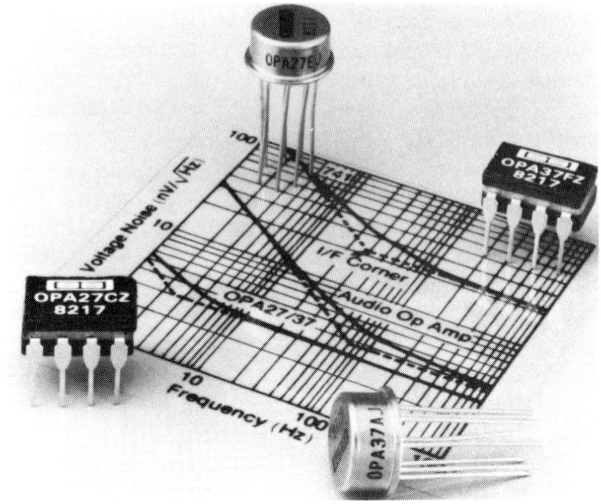

Figure O.3 Examples of precision operational amplifiers (OP AMP) (*courtesy* Burr-Brown International Ltd)

OPM Operations Per Minute. The number of defined actions executed in one minute.

OPR Optical Pattern Recognition. A technique in which an optical scanning device is used to convert scanned information into digital (D,4) signals. The method is normally used for the computer recognition of patterns by comparison to a known reference, for classification, identification or sorting.

OR (1) Operational (or Operations) Research. Scientific methods applied to the solving of problems that concern the allocation of resources. They make use of mathematical models and analytical methods to provide a logical basis for solving operational problems as well as for making sound predictions

MIL-STD-806B	BS 3939 ANSI Y.32.14-1973 IEC 117	Old BS 3939	Sometimes used symbol	Truth table

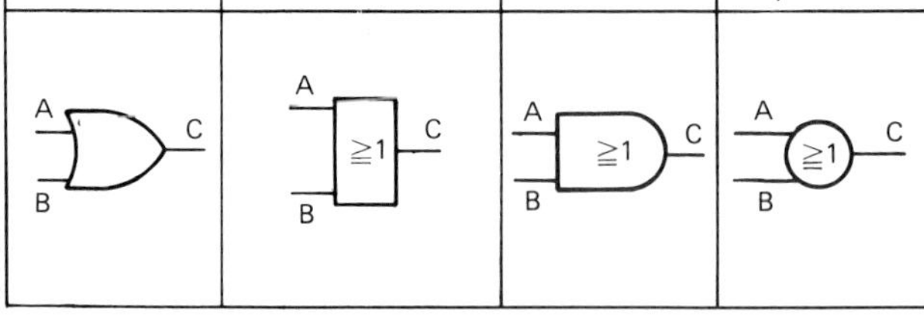

A	B	C
0	0	0
0	1	1
1	0	1
1	1	1

Figure O.4 The OR gate. In practice the MIL-STD-806B symbol is the one most likely to be met, although the other logic symbols may be encountered

and decisions that have been designed for and are used by management.
(2) OR gate. This logic element has one output (O,2) and two or more inputs. Its output will stand at its defined logic 1 state if one or more of its inputs stand at their defined logic 1 state. *See Figure O.4.*

ORACLE The name given by the Independent Broadcasting Authority (IBA) for its broadcast information service. It consists of the transmission of information in a coded form within the standard television signal, thus providing text display (D,6) or pictorial matter on the screen of receivers that have the necessary circuitry for reception (teletext systems).

ORBIT On-line Real-time Branch Information. A computer language developed by International Business Machines (IBM) to search stores of information held on computer files.

ORGALIME ORGAnization for LIaison between electrical and Mechanical Engineering industries. An association that represents the national trade organizations for electrical and mechanical engineering, of the European Economic Community (EEC) and European Free Trade Association (EFTA) countries.

ORSA Operations Research Society of America. A body that is interested in extending the computer far beyond automatic and routine operations, into decision making and other complex processes of management.

OS Operating System. (a) The organized collection of procedures and techniques that are required for the operation of a computer. (b) That part of computer software necessary to simplify administrative procedures for input/output (I/O), data-conversion and other routines.

O/S Operating System. *See* OS.

OSCAR Orbital Satellite-Carrying Amateur Radio. The name of a number of satellites launched by The Radio Amateur Satellite Corporation (AMSAT).

OSD Optical Scanning Device. A technique in which a device scans graphics or text and converts the scanned information into digital (D,4) signals for computer processing.

OSDM Optical Space-Division Multiplexing. Bundles of independent fibres that provide optical paths for independent channels (CHNL(d)). In practice this is a single cable with overall sheathing and strengthening made from a collection of fibres. *See* SDM, space-division multiplexing.

OSI Open System Interconnection. A term used generally to describe the technique employed for the construction and connection of a computer network, in which the types of interconnected computer systems employed are different. Proposed specifications of such a system have been drafted by the International Standards Organisation (ISO).

OTD Optical Time Domain. A distance, measured in time, from a light source to a reflective surface and back again. The concept is used for checking fibre-optic (FO) cables with an OTD reflectometer. The reflectometer launches a light pulse into a fibre-optic cable and measures the time taken for its reflection to return, thus indicating continuity or discontinuity caused by cracks, fractures or breaks.

OTP Office of Telecommunications Policy. An executive office of the US President that has no statutory powers of regulation but develops and recommends public policy in the area of telecommunications.

OTS Orbital Test Satellite. A programme started in 1977 by the European Space Agency for experimental communication satellites (COMSAT).

OTS II Orbital Test Satellite II. A geostationary satellite launched in 1978 as a test station for the equipment, techniques and systems intended for further use by the European fixed-service satellite system (ECS) of the 1980s. OTS II has the capacity to deal with 6000 telephone calls and two television (TV(a)) channels (CHNL(g)) simultaneously. OTS I was never brought into commission due to failures during its launch.

OV OVerflow. An overcapacity, for instance the generation in an arithmetic operation of a quantity beyond the capacity of the memory store or location to receive it. Fields that are longer than field locations will overflow, e.g. a 10-digit product will overflow an 8-digit accumulator (ACC).

OVD Optical Video Disk. A disk having on its surface a spiral or concentric circles of recorded digital (D,4) data at a high packing density (D,3). The information is recorded in spots using a laser (LASER) beam and is read by reflective means using a low intensity laser beam. Using a suitable machine the disk can be used to provide television (TV(a)) programming playback.

OVP Over-Voltage Protection. The action of defending a circuit (CCT) from voltages (V,2) larger than predcterinined requirements. *See* OVPU, over voltage protection unit.

OVPO Outside Vapour-Phase Oxidation. A chemical vapour-phase oxidation (CVPO) process that is used for the production of optical fibres (OF). Dopants are burnt with oxygen (O,1) and gas to form an oxide stream that is deposited onto the outside surface of a rotating glass rod.

OVPU Over-Voltage Protection Unit. A circuit (CCT) or device that comes into operation when the supply voltage (V,2) in a particular circuit rises above a predetermined value. Such units are used to protect computers and delicate equipment from short-duration transients (spikes) in supply voltages.

OWF Optimum Working Frequency. In high frequency (HF) transmissions, an OWF is a frequency that is lower than the maximum useable frequency (MUF) by an amount sufficient to allow for expected variations in the propagation conditions that may occur during the working period of operation.

P

p (1) Pico. A decimal prefix commonly used in association with a base unit in the Système International d'Unités (SI) system of units indicating a multiplication factor of 10^{-12} of that unit. For example, one picofarad (pF) is equal to 0.000 000 000 001 farad.
(2) P-channel. A metal oxide semiconductor (MOS) circuit in which the conducting path is formed as a p-doped (p-type) semiconductor. This type of channel (CHNL(h)) can also be found in junction field-effect transistors (JUGFETs).

P (1) Peta. A decimal prefix used in association with a base unit in the Système International d'Unités (SI) system of units indicating a multiplication factor of 10^{15} of that unit. For example, one petatonne, Pt, is equal to 1 000 000 000 000 000 tonnes.
(2) P band. A band of microwave (μw) frequencies that range from 0.225 to 0.390 gigahertz.
(3) Power. The rate at which work is done or energy is expended; the unit is the watt (W,2). In a direct current (DC) circuit or device, W = VI, where V is the potential difference (PD) in volts (V,1) and I the current (I) in amperes (A) *See also* PWR.

Pa PAscal. The unit of pressure in the Système International d'Unités (SI) system of units. It can be defined as the pressure that results when one newton (N) acts uniformly over an area of one square metre (m).

PA (1) Public Address (system). Usually consisting of one or more microphones (MIKE), power amplifiers (AMP) and several loudspeakers, PA systems are designed to reproduce sounds, most often speech or music, so they become audible to a large gathering of people. They are used for airport announcements, in railway stations, open-air meetings or large halls.

(2) Paper Advance. Usually a key or button provided on a printer that moves the paper or printout forward independently of the action of the printing head or impressing device.

PABX Private Automatic Branch eXchange. A privately owned automatic branch telephone exchange that provides access to and from the public telephone network without operator intervention.

PAC Program Address Counter. A sequential counter that is able to keep track of the location of the next instruction to be executed from a program memory.

PACC Product Administration and Contract Control. A particular method of control employed by a data management system in a business environment.

PACVD Plasma-Activated Chemical Vapour Deposition. A method for producing graded index (GI) optical fibre (OF) using a form of the chemical vapour deposition (CVD) process. Thin layers of different indices are deposited on the inner wall of a glass tube, a chemical vapour flows through the tube and the formation of oxides is stimulated by means of plasma generated by a microwave (μW) resonant cavity.

PAD (1) Packet Assembly/Disassembly. Provided in a packet-switching exchange, a PAD facilitates communication between two terminals of a different nature. It has two basic functions, firstly to assemble received data into packets for onward transmission and secondly to disassemble these packets and transmit to various other terminals.

(2) Packet Assembler/Disassembler. A unit that provides a packet assembly/disassembly (PAD,1) facility.

PAL (1) Phase Alternation Line. A system used in the UK, most of Europe, and elsewhere, for 625-line colour television (TV(a)) transmission. Signals are much less susceptible to transmission problems than in other systems, as the technique that is used inverts the phase of one of the two chrominance signal components every alternate line.

(2) Programmable Array Logic. A variant of programmable logic array (PLA) which is optimized in its application, providing more inputs (I/P), outputs (O/P) or functions.

PAL D Phase Alternation Line Delay. A method of decoding signals used in the Phase Alternation Line (PAL) colour television (TV) system, that reduces colour errors.

PALLNIC PALLadium-NICkel. A palladium/nickel alloy of between 65% and 85% palladium content, with a colour and finish similar to that of bright stainless steel. It was developed to fulfil needs within the electronics industry to replace gold in connectors and printed circuit board (PCB) edge connectors.

PAL S Phase Alternation Line Simple. A process of decoding signals used in the Phase Alternation Line (PAL) colour television (TV) system, that requires the observer by sight to optically integrate any switched, line-by-line colour errors.

PAM Pulse Amplitude Modulation. A form of modulation used in communications, where the amplitude of the pulse carrier is varied in accordance with successive samples of the modulating signal.

PANDA Prestel Advanced Network Design Architecture. A particular design of construction for interconnecting Prestel devices.

PAR Precision Approach Radar. An aid to air traffic control, used at an airport to provide accurate information concerning the location of incoming aircraft in the local vicinity. It is usually used in association with an airport surveillance radar (ASR).

PARD Periodic And Random Deviation. A term recommended by the International Electrotechnical Commission (IEC) as the specification for ripple and noise.

PASCAL (1) A computer language designed originally to enable the teaching of programming as a systematic discipline and to allow those being taught to do systems programming. It is based on an algorithmic language (ALGOL). First developed at the Zurich Technical Institute in the late 1960s it emphasizes aspects of structured programming.
(2) Program Applique à la Sélection et la Compilation Automatique de la Literature. Compiled by the Centre National de la Recherche Scientifique in France, this is a very large store of information held on computer files. It covers a wide range of different subjects.

PATCH A colloquial expression (a) normally employed to describe a non-standard or unusual correction to a computer program that corrects a bug or unexpected function, (b) employed to describe an extra connection or link that was not intended at the design stage but enables a circuit (CCT) or a printed circuit board (PCB) to function. (c) In communications, a non-standard or unusual route connecting two parties.

PATRICIA Practical Algorithm To Receive Information Coded In Alphanumeric. A set of simple procedures or rules for receiving alphabetical and numerical information.

PATSY Programmer's Automatic Test SYstem. A procedure that assists people who prepare programs for a computer, in debugging routines.

PAU Pattern Articulation Unit. A microprocessor (MP) that is capable of converting graphics images into a stream of data or information. It can also reverse the process, i.e. reconstruct graphics images from converted information.

PAX Private Automatic eXchange. A dialled telephone exchange that provides a private service within an organization, but does not allow calls to be transmitted to or from the public telephone network.

PB Peripheral Buffer. An input/output (I/O) buffer situated within the memory store of a peripheral device. It permits data-word transfers to and from the computer memory without the need of main program attention.

P band P band microwave (μW) frequencies occur between 0.225 and 0.390 gigahertz (GHz) and form part of a larger continuous range of frequencies covering 0.225–100 gigahertz which are usually subdivided into bands designated by letters. The bands are commonly used subdivisions but, however, are not internationally agreed. P band frequencies have wavelengths of between 133.3 and 76.9 cm.

PBJ Paper-Braided Jute. An insulation material for cables.

PBX Private Branch eXchange. A private telephone exchange located on a subscriber's premises that has access to the public telephone network. The term can be applied to automatic or manual exchanges, although in the UK an automatic exchange is referred to as a PABX and a manual exchange is referred to as a PMBX.

PC (1) Printed Circuit. A circuit (CCT) in which all the interconnecting wires have been replaced by conductive strips, normally copper. These strips have to be plated, printed or etched onto an insulating baseboard. *See* PCB, printed circuit board.

(2) Personal Computer. A small low-cost computer (relative to standard and mini systems) available as ready-built units or in kit form. Applications include professional business, small business, hobbyists and home, the individual software being orientated towards single user applications. The two main categories are (a) those that have their own cathode-ray tube (CRT) display (D,6) and cassette or floppy disk (FD,1) facilities for program storage and exchange; and (b) those that rely on a domestic television (TV(c)) set as an output (O,2) display and an audio cassette for program storage and exchange. Both are able to accommodate peripheral devices such as standard or purpose designed printers.

(3) Portable Computer. A hand-held micro or personal computer (PC,2). Typically these devices can accommodate add-on printers or cassette units.

(4) Program Counter. A processor register that is automatically

advanced, usually by 2, every time the processor retrieves an instruction. This information is then used in connection with operating modes.

P-C Processor-Controller. With reference to computer systems, a processor-controller can be used for direct control, data analysis, supervisory control or editing.

PCB Printed Circuit Board. A circuit (CCT) or part of a circuit formed on an insulating board or card, in which the conducting interconnecting paths have been printed or etched. PCBs may be single-sided, double-sided or multilayer for high-density applications.

PCB PS Printed Circuit Board Power Supply. A power supply unit (PSU) mounted on a single printed circuit board (PCB) or card.

PC DOS Personal Computer Disk Operating System. Developed the Microsoft Corporation for IBM's (see IBM) then new personal computer (PC,2), PC DOS has rapidly become a standard operating system (OS) for program development. See disk operating system (DOS).

PCI (1) Panel Call Indicator. Relays or switches at a terminating office that register the detection of a direct current (DC) pulsing system. In this system, information digits (DGTs) are each transmitted as a series of four marginal, polarized pulses.

(2) Program Check Interruption. Caused by unusual conditions experienced in a computer program, this interruption is usually due to an incorrect operand.

(3) Process Control Interface. Manufactured in the form of a circuit on a board, this device can provide computers with an interface to a variety of electromechanical devices such as relays, switches, motors, contacters, thermostats and various other devices.

PCM (1) Pulse Code Modulation. A form of modulation used in communications where the modulating signal is sampled and the sample quantized and coded so that each element of the information consists of one or more binary digits (BITs). In a communication system this type of signal modulation allows noise-independent transmission, as the signal is made up of a rapid sequence of pulses which have the same amplitude, arranged in binary groups. These correspond to numerical groups which represent the amplitude values needed to recreate the slope of a signal, or the pulses can represent data bits.

(2) Punched (Punch) Card Machine. Widely used items of equipment that punch, read, analyse or deal in any way with, punched cards. These include sorters, collators and tabulators.

(3) Plug Compatible Manufacturer. A manufacturer of

peripheral equipment that produces items fully compatible with a particular computer, the computer being made by another manufacturer. The devices simply plug into the computer.

PCMI Photo-Chromic Micro Image. An ultrafiche having extremely small images such that over 3000 pages can be held on an A6 card. It is similar to a microfiche, but requires much higher standards of control during production.

PCP (1) Plug Compatible Peripherals. Peripherals to a computer, which can include main memory, that simply plug in. The PCPs are made by a different manufacturer than that of the computer.

(2) Primary Control Program. A particular sequence of instructions that prescribe the series of steps to be taken by a computer system, the system having been designed to perform one task or job at a time.

(3) PolyChloroPrene. A material used for cable sheathing which is heat resistant, oil resistant, flame retardant (HOFR). Also known as Neoprene.

PCS (1) Personal Computing System. A relatively low cost portable computer that has software designed for single-user applications, personally controllable and easily used. There are several classifications of systems, which include home, hobbyist, professional and small business.

(2) Plastics-Clad Silicon. Optical fibre (OF) with a core of pure silica glass covered by an overall plastics cladding. Its loss and dispersion are normally higher than those of other types of fibre-optic (FO) material.

(3) Print Contrast Signal. A value of a measured difference between the contrast of a character that has been printed and the contrast of the paper on which that character appears.

(4) Punched-Card System. A system for encoding information onto cards in the form of holes. The pattern of positions punched for any given character is determined by the particular code employed by the card punch.

PD (1) Potential Difference. In the Système International d'Unités (SI) system of units the unit of potential difference is the volt (V,1). It can be defined as the line integral of the field strength along a freely chosen path between two points in an electric field that is produced in the absence of electromagnetic induction.

(2) Photo Detector. A light-detecting device that can be used for communications by extracting information from an optical carrier.

(3) Pen Down. A command in the Logo (LOGO) computer programming language that reserves the Pen Up command (*see* PU).

PDA Problem Determination Aid. A device or routine that assists the procedure which is to be adhered to in order to determine the cause of an error.

PDD Post Dialling Delay. A time interval that occurs between the finish of dialling on a telephone receiver and the reception of a return signal, which could be either a ringing tone or a busy tone (BT,2).

PDL Print Definition Language. Used by a computer to define printer characteristics including lines, columns, page and overflow print conditions.

PDM Pulse Duration Modulation. A particular form of pulse modulation (PM) in which the duration of each pulse in a pulse train is varied according to the instantaneous value of the modulating wave. The modulating signal can vary when the leading edge, the trailing edge or both edges of each pulse occurs.

PDP Programmed Data Processor. A wide range of computer systems from large mainframes to minicomputers, manufactured by the Digital Equipment Corporation.

PDS Partitioned Data Set. A computer file that is divided into subfiles which are able to be referenced as a group or individually.

PDU Pilot Display Unit. A display (D,6) unit mounted in a cockpit on which information is projected in the pilot's line of sight and superimposed over his view of the real world. *See* HUD, head-up display.

PE Phase Encoding. A particular technique employed for encoding information onto a magnetic tape (MAGTAPE).

PEC (1) Photo-Electric Cell. A light-sensitive device comprising two electrodes in a glass bulb that can be evacuated or filled with an inert gas. The unit provides an electrical output (O,2) directly dependent upon the amount of light falling on its input electrode.

The term more commonly used is 'photocells' and these are often made from semiconductor materials. There are several different types of photocell which can be either photo conductive, or photovoltaic, or photo-emissive.

(2) Panel Electronic Circuit. A printed wiring panel that has components mounted upon it.

PEL Picture ELement. The smallest detail in a picture area, comprising a television (TV(a)) facsimile (FAX) or other transmitted picture, that can be resolved by the scanning system without loss of amplitude; also known as 'pixel'. *See* PIXEL. *See Figure P.1*.

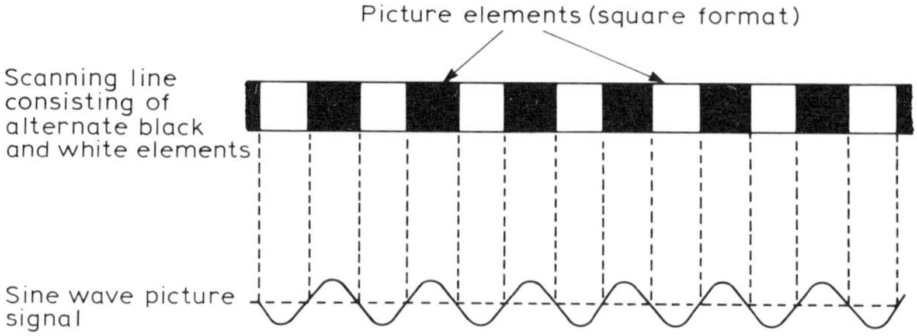

Figure P.1 A picture signal corresponding to a row of alternate black and white picture elements (PEL)

PEMS Porcelain Enamelled Metal Substrates. Replacements for printed circuit boards (PCBs) and thick film circuits on ceramic. Circuit boards using PEM substrates act as their own heat sink/electrical ground plane; are not affected by severe conditions of humidity, temperature and thermal or physical shock; and may be formed into a wide variety of shapes to meet specific applications.

PEP Peak Envelope Power. The mean power supplied to an antenna during one complete radio frequency (RF) cycle at the peak of the modulation envelope.

PEPE Parallel Elements of Processing Ensemble. An early method of executing a number of processes in a computer, in parallel.

PERIPHERAL PERIPHERAL device or equipment. An auxiliary unit or piece of equipment of a computer system. The devices normally perform input/output (I/O) operations, such as disk drives or paper tape readers (PTR).

PERMALLOY PERMeable ALLOY. An alloy with a high magnetic permeability (at low values of magnetic flux density), low coercivity, and therefore low hysteresis loss. It contains 78.5% nickel and 21.5% iron, although varieties are produced by adding copper, cobalt, manganese and other metals.

PERT Program Evaluation and Review Technique. A computer program that provides a method of analysis and reporting for overall project administrators. Areas that impose the greatest time restriction for completion of a project as well as those with excess time for completion can be highlighted. Similar to critical path method (CPM).

PET Personal Electronic Transaction computer. A popular personal home computer, based on the 6502 8-binary digit (BIT) microprocessor (MP) manufactured by Commodore Business Machines.

pF PicoFarad. One millionth of a millionth of a farad (F), that is 0.000 000 000 001 or 10^{-12} farads.

PF (1) Pulse Frequency. The number of complete pulses measured during a particular time interval.

(2) Page Footing. A record of information or data situated and printed at the bottom or foot of a page(s) of a document.

(3) Packing Fraction (or Packing Factor). A ratio of the active core total cross-sectional area of a bundle of optical fibres (OF) or cable cores, to the total cross-sectional area of the bundle. The packing fraction can be expressed as $PF = N(A/D)^2$ where N is the number of cores, A is the diameter of each core and D is the diameter of the whole bundle or assembly.

(4) Power Factor. A comparison of average power dissipation against the apparent power in an alternating current (AC,1) network or part of that network. It is expressed as a ratio.

PFM Pulse Frequency Modulation. A form of pulse modulation (PM) in which the pulse repetition rate is varied.

PFR Power Fail Recovery (or Power Fail Restart). A system containing automatic switching and charging circuits, designed to protect computer systems by detecting a drop in input (I/P) voltage (V,2), which signals an imminent power failure. This provides several tens of milliseconds (ms) during which all the memory locations can be maintained by a battery backed-up memory, or preserved in a non-volatile memory (NVM) and then automatically restarted when the power is restored.

PGA Pin Grid Array. A mounting technique employing a high density package for very-large-scale integration (VLSI) integrated circuits (IC).

PGD Planar Gas Discharge. A type of display device which produces highly visible yellow-orange light on a high-contrast black background.

PGS Program Generation System. Allowing the user to select a particular area of memory and output (O,2) in a particular format, this system will then load programs into memory. It can load self-produced programs and also others produced elsewhere.

pH Percentage Hydrogen. A measure of acidity, expressed as the common logarithm of the reciprocal of concentration. It indicates the concentration of hydrogen ions in a solution.

PH Page Heading. A record of information or data that is situated and printed at the top or head of a page or pages in a document.

pi PI-network (or π-network). A particular type of network or circuit (CCT) comprising two equal shunt arms with a series arm between them.

PI (1) Programmed Interrupt. Often used as a fast method of transferring control to an error routine, a PI is caused deliberately by a computer program (as opposed to one caused by an external event).

(2) Paper Insulated. Generally refers to power cables that use insulating paper having a density of between 0.7 and 0.9 g/cm^2 when dry. The paper is impregnated with insulating oil to improve its dielectric strength.

PIA Peripheral Interface Adaptor. A device that will permit parallel orientated peripherals to be connected to various microprocessors (MP) usually without any additional circuits.

PIC Priority Interrupt Control. A special chip (CHIP) in a microcomputer that manages several, normally eight, external attention signals sent from an input/output (I/O) device or chip to the microprocessor unit (MPU) to obtain service.

PICTEL PICture TELephone. A name used by the US General Telephone and Electronics Company to describe its experimental video telephone that permits the user to see as well as talk with another party.

PID Peripheral Interface Device. Interface printed circuits, compatible with a set of standard peripherals, that can all be plugged into a chassis assembly. It may sometimes be a separate unit.

PIL Precision-In-Line. A colour picture tube employing a shadow mask near the screen in which there are vertical slits. The screen has vertical stripes of red, blue or green phosphor dots. Three electron guns or one that fires three beams are used and the mask ensures that any one strip is hidden from two of the beams.

PILOT An interactive computer-aided instruction (CAI) computer programming language developed by John Starkweather at the University of California.

PIM Pulse Interval Modulation. A form of pulse modulation (PM) in which constant amplitude pulses of constant width have their spacing varied.

pin or **p-i-n** Positive-Intrinsic-Negative. A junction semiconductor diode (D,5) with an area of semiconductor material nearing intrinsic conductivity (i-type) between the positive p-doped semiconductor material (p-type) and the negative n-doped semiconductor material (n-type). At low frequencies the diode acts similarly to a p-n semiconductor junction (p-n), but at high frequencies it exhibits variable resistance (R,1) and is suitable for impact ionization avalanche transit time (IMPATT) operation. P-i-n diodes are used as photodetectors (PD).

PIN (1) *See* pin.

(2) Personal Identification Number. A number used in conjunction with a banker's card to activate an automatic teller machine (ATM).

PINFET Positive-Intrinsic-Negative Field-Effect Transistor. This device is intended for long wavelength, high-bit-rate fibre optic communications (FOC). It combines the high sensitivity of a positive-intrinsic-negative (PIN) photodiode with the low noise performance of a field-effect transistor using a hybrid construction with the chips (CHIP) positioned as close together as practicable to reduce parasitic circuit (CCT) impedances.

PIO (1) Parallel Input/Output. An interface that permits a computer to input (I/P) and output (O,2) data bits simultaneously to and from external parallel devices, such as a visual display unit (VDU) or a keyboard (KB). It is a circuit (CCT) in a computer system or a separate unit that is bus compatible with various microcomputers and peripheral devices. Most are programmable to operate in static, clocked (CLK) or discrete modes.

(2) Programmable Input/Output chip. An input/output (I/O) chip (CHIP), in most cases an 8-binary digit (BIT) interface chip that multiplexes (MUX) a connection to a data bus, which is capable of transferring data to and from the central processor unit (CPU) and peripheral devices, into two or more ports.

PIP (1) Peripheral Interchange Program. A series of computer programs that transfer information from one device to another.

(2) A significant intensification or deflection of the spot on a cathode ray tube (CRT), that is providing a display (D,6) for identification or calibration. Normally it is applied to the peaked pattern of a cursor signal or device.

PIRS Personal Information Retrieval System. A software package used together with a mini or microcomputer to order and index information for storage on a magnetic medium.

PIT Programmable Interval Timer. A chip (CHIP) that has a separate timing clock (CLK) and a number of special memory storage locations. These are used independently of the microprocessor unit (MPU) to count time for real-time applications. They can store the time-elapsed or generated signals, at the completion of a time period.

PIU Programmable Interface Unit. A programmable large scale integration (LSI) device that is used to eliminate the hardwire logic often needed for communication interfaces, between processors and various other system components.

PIV Peak Inverse Voltage. The maximum instantaneous voltage (V,2) across the input terminals of a diode (D,5) in a rectifying circuit, when the diode is non-conductive.

PIXEL PICTure ELement. A small rectangular division of a video screen. More, and smaller, rectangles are necessary to achieve higher resolution. Therefore a sharper picture requires more computer memory to store the pixels. *See* PEL.

PL/1 Programming Language ONE. Developed by International Business Machines (IBM) in the mid-1960s, this is a high level computer language employed in a varied range of scientific and commercial applications. A version now exists for microcomputers.

PLA Programmable Logic Array. An orderly arrangement of logical AND (*see* AND) functions and logical OR (*see* OR) functions, which are constructed using metal-oxide semiconductor (MOS) or bipolar circuits. They are used as alternatives to random-access memories (RAM) which use standard logic networks programmed to perform a specific function.

PLATO Programmed Logic for Automatic Teaching Operations. A computer language and also a system of computer-aided instruction (CAI).

PLC Programmable Logic Controller. A self-contained system or unit that includes all the elements of computer-control, arithmetic and logic functions, memory, input/output (I/O) and the required software to make them implement the desired control function. Such a device is therefore capable of being programmed to obtain particular desired logic control functions.

PLCC Plastic Leaded Chip Carrier. PLCCs were developed for use on epoxy glass fibre laminated boards. They have flexible leads similar to small outline (SO) packages and can thus compensate for the mechanical stress resulting from the different thermal coefficient of expansion of board and package. They have 'J'-shape leads which are bent down under the component, *see Figure P.2*.

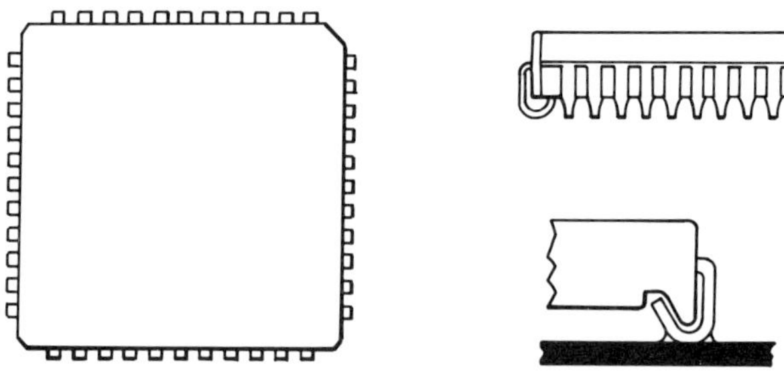

Figure P.2 Detail of a typical plastic leaded chip carrier (PLCC) packaged, surface mounted device (SMD)

PLD Programmable Logic Device. A logic device that constitutes just a part of a complete computing system, such as erasable programmable read-only memories (EPROM), read-only memories (ROM) and field-programmable logic arrays (FPLA).

PLL Phase Locked Loop. A phase comparator and a voltage controlled oscillator (VCO) linked so that the phase oscillator frequency can precisely monitor that of a phase-modulated signal or applied frequency. If they differ in phase a control voltage (V,2) is developed which, when filtered, changes the frequency to reduce error.

PL/M Programming Language for Microprocessor. A high-level programming language that permits users to employ a notation with which they are already familiar. It was developed by Integrated Electronics (INTEL), for use on its microcomputer systems, from a dialect of programming language one (PL/1).

PLM Pulse Length Modulation. A form of pulse modulation (PM,2) in which the duration of each pulse in a pulse train is varied according to the instantaneous value of the modulating wave. The modulating signal can vary when the leading edge, the trailing edge, or both edges of each pulse occurs.

PLO Phase Locked Oscillator. A phase locked loop (PLL) circuit that is used for precise data recovery in floppy disk drive controllers (FDC). It stabilizes the separated data and clock (CLK) binary digits (BITs).

PLP Presentation Level Protocol. An agreed viewdata standard between Canada and the USA for the whole of North America.

PM (1) Phase Modulation. A method of modifying a sine wave signal to make it carry information, in which the carrier has its phase changed according to the information to be transmitted. In operation the angle relative to an unmodulated carrier angle is varied in accordance with the instantaneous value of the amplitude of the modulating signal. For digital transmission, 2, 4, or 8 different phases can be employed.

(2) Pulse Modulation. A method of transmitting information by modulating a pulsed or intermittent carrier. Pulse width, count, position, phase and/or amplitude may be the varied characteristic.

(3) Post Mortem. (a) An analysis undertaken of an event or operation after the completion of that event or operation. (b) In a computer system, a check routine that provides information concerning the contents of all or a particular section of storage after a problem has ceased to exist. It assists in locating a program error or machine malfunction.

PMBX Private Manual Branch eXchange. A private telephone exchange located on a subscriber's premises, operated manually and providing access to the public telephone network.

PMD Post Mortem Dump. The transfer of the contents of the computer memory instantaneously to a peripheral unit, performed at the end of a machine run. It is used to provide information for locating, diagnosing, and correcting errors.

PML Programmable Macro Logic. The architecture of conventional logic can be replaced by PML with its logically equivalent structure (NAND-NAND. *See* NAND). The central programmable NAND array structure provides maximum interconnect flexibility.

PMOS P-type (or channel) Metal-Oxide Semiconductor. The earliest of the large-scale integration (LSI) chips (CHIPs). It provides excellent component density, but is slower in performance than N-channel metal-oxide semiconductor (NMOS) devices. The operating speed is of a few megahertz, the device requires two power supplies and is not transistor-transistor logic (TTL) compatible.

PMR Private Mobile Radio. A general range of land mobile radios, which extend from licensed taxi radios to hand portables. The growth of users continues unabated and is compounded by the newer services such as cellular mobile telephones and cordless telephones (CT).

pn or **p-n** P-doped N-doped junction. A junction or transition (depletion layer) between a p-doped semiconductor (p-type) and an n-doped semiconductor (n-type) having rectifier properties. It allows only a small current to flow due to free electrons and holes being produced near the junction by thermal agitation. *See Figure P.3.*

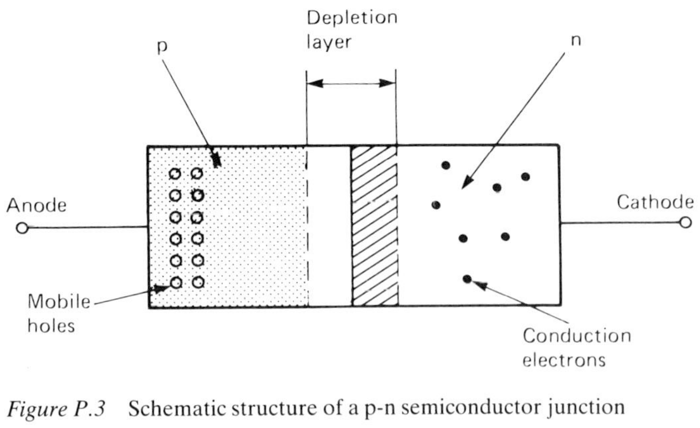

Figure P.3 Schematic structure of a p-n semiconductor junction

PN (1) Punch oNly. The function of a paper tape punch (PTP,2) which is a device that punches blank paper tape, but is unable to sense or translate the holes.

(2) Paging Network. Used within a public switched telephone network (PSTN) and a cellular radio network (CRN), the PN sends a paging signal to a universal portable telephone (UPT) in areas where voice communications are not possible.

PNA Project Network Analysis. *See* PNT.

pnip or **p-n-i-p** P-doped N-doped Intrinsic P-doped transistor. A transistor that is constructed with an intrinsic (undoped) layer of material between the base and collector in order to extend the high-frequency range.

pnp or **p-n-p** P-doped N-doped P-doped transistor. Consisting of two p-doped regions (p-type) with an n-doped region (n-type) separating them, this type of transistor is characterized by the flow of 'holes' as majority carriers rather than the more mobile electrons.

pnpn or **p-n-p-n** P-doped N-doped P-doped N-doped semiconductor device. Fabricated in silicon, a device that consists of alternating p-doped semiconductor material (p-type) and n-doped semiconductor material (n-type) with at least three p-n junctions (p-n). A basic pnpn structure with no gate electrodes is termed a four-layer diode (D,5) or Shockley diode. It has a bistable current-voltage characteristic and is used for power-switching purposes, as shown in *Figure P.4*.

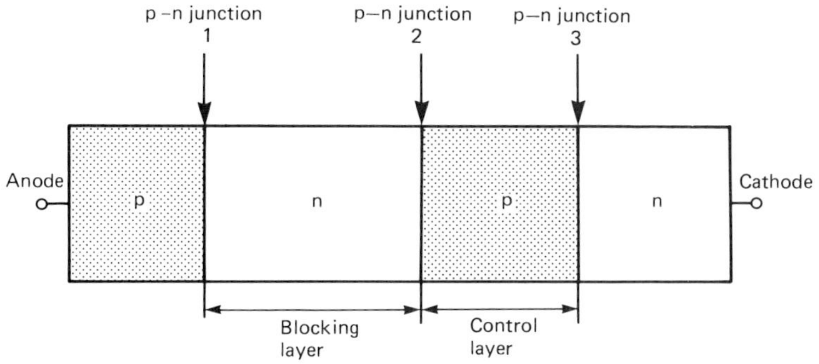

Figure P.4 A basic pnpn structure with no gate electrodes; known as a four-layer diode, or Shockley diode

PNT Project Network Techniques. A group of techniques that considers the logical inter-relationship of all project activities. It is used for the description, analysis, planning and control of projects. The group includes techniques concerned with time, resources, costs and other uncertain factors.

PO Post Office. The British Post Office, formerly known as the General Post Office (GPO). The PO has set many standards for the production of electronic devices and equipment.

POA Point Of Action. Describes displays (D,6) and readouts sited where processes such as manufacturing actually take place. They are characterized by housings which can withstand environmental hazards such as impact and the ingress of dust and water. They have bright wide-angle characters that are often laid out in a single line, so that messages are received by operators without the need to look closely at visual display units (VDUs).

POCSAG Post Office Code Standardisation Advisory Group. A group that provides consultation on standards, e.g. those for national wide area paging receivers.

POD Point of Origin Device. PODs in a computer system can be teletypewriter units, consoles, meters, optical character readers (OCR), video displays (CRT) and any of a number of input instruments able to send machine-sensible signals, which are connected directly to the central processor or one of the peripheral processors in a time-shared system (TSS,1).

POKE A command statement used in Beginner's All-purpose Symbolic Instruction Code (BASIC). This self-modifying feature will cause a computer or microcomputer to change a section of its own program in a manner the programmer commands.

POL Problem Orientated Language. A computer language designed for the solutions of a given class of problems. Although elaborate it should be easily understood and usually allows the solution to a problem to be expressed in the same terms as the problem.

POLISH POLISH notation. Designed in 1929 by a Polish logician, Jan Lukasiewicz, this is a representative notation in which each operator precedes the operand upon which it operates.

POLYSILICON POLYcrystalline SILICON. A polycrystalline form of silicon generally used as the gate electrode in silicon gate metal-oxide semiconductors (MOS), integrated circuits (IC) and charge-coupled devices (CCD). In this particular application the silicon exhibits metallic properties, as it degenerates due to high doping concentrations.

POP A colloquial expression for the retrieval of information from a push down stack, in order to place it in a particular memory location or register (as opposed to entering information which is pushed. *See* PUSH).

POS Point-Of-Sale. Refers to computer systems that are

employed in many parts of retail operations. They can be used for credit authorization, credit verification, stock control or for the transfer of funds.

POSH Permuted On Subject Headings. An index that lists all the subject headings of a document's title so that each heading appears, in turn, as the leading word followed by the remaining words.

POT POTentiometer. Normally a circular or disk-type of resistor (R,2) that has two fixed terminals and another third terminal which is connected to a movable contact arm. It is most commonly found as a volume or tone control on audio and television (TV(c)) equipment.

POTS Plain Old Telephone Service. A term used, mainly in the USA, by the telephone industry for a conventional telephone service.

PP Peripheral Processor. Used in association with one or more pieces of computer peripheral equipment (PERIPHERAL), this processor helps the peripheral to function semi-autonomously, not under the control of the central processor unit (CPU).

P/P Peak-to-Peak. The voltage (V,2) difference between the maximum (MAX) and minimum (MIN) values (typically positive and negative) of an alternating signal.

PPI (1) Plan Position Indicator. A type of presentation used in a radar receiver. The signal received from the target is shown as a bright spot on a screen. The distance and bearing of the target are given by the polar co-ordinates of the spot with respect to the centre of the screen.
(2) Programmable Peripheral Interface adaptor. A peripheral interface adaptor that can provide control over the connection of parallel orientated peripherals to various microprocessors (MP).

PPM (1) Pulse Position Modulation. A form of pulse modulation (PM) in which the positions, in time, of pulses are varied without their duration being modified.
(2) Peak Programme Meter. An instrument specifically designed to measure the peak values of audio signals within a predetermined period. It is able to respond very quickly to increases in audio signal amplitude, thus enabling accurate measurement of peaks, but takes appreciably longer for indicators to return to zero.
(3) Period Permanent Magnet. A magnet constructed from alternate permanent magnetic rings and soft-iron pole pieces that is able to provide the necessary axial magnetic field required in travelling-wave tubes and other similar devices.

pps (1) Pulses Per Second. The number of pulses which occur per second (s) experienced at a particular point in a transmission medium of free space, or at a point in a circuit (CCT), device or system.

(2) Parallel Processing System. A system including a central processor unit (CPU), random-access memories (RAMs), read-only memories (ROMs), and clock (CLK) circuits, with circuits (CCT) that can be used to build equipment requirements into digital (D,4) data processing (DP,1) equipment. Such systems can be found in calculators, general-purpose data processors, process controllers and cash registers.

PPSU Programmable Power Supply Unit. A particular type of laboratory power supply (LabPS) unit that has an internal memory which can be programmed with the value of voltages (V,2), each with an associated current-limit and dwell time. A typical unit has an IEEE-488 (*see* IEEE-488) interface and a memory that once loaded can be stepped through, run continuously, or run in a single slot mode. Such features make PPSU devices ideal for use with automatic test equipment (ATE).

PR PRefix. A number, syllable or word, placed at the beginning of another number, syllable or word, which together form a compound that has a precise value or meaning.

PRE PREfix. *See* PR.

PREAMP PREAMPlifier. An amplifier used to amplify received signals before they are fed into the rest of the system. It is used in a sound reproduction system such as a radio receiver, in an antenna system, and in television (TV(a)) cameras.

PRECIS (1) PREserved Context Index System. A system of indexing information in which various groups or strings of items are arranged with respect to their linguistic function. This permits a computer to use the information strings so that the user can find a term which is of primary interest.

(2) PRE-Coordinate Index System. A system of indexing documents in which terms are combined as the index is being compiled to provide the entries.

PRISM Personal Records Information System Management. A computerized personal information retrieval system (PIRS) used by the UK Civil Service.

PRF Pulse Repetition Frequency. The number of pulses measured during a particular time interval, which are independent of that interval. The term is normally used for a regular train of pulses, i.e. the number of pulses per unit time is constant. With reference to computers, the PRF is the number of electric pulses experienced in a particular unit of time, measured at one point within the system.

PROLOG PROgramming in LOGic. A logical programming language, popular in artificial intelligence, that is being used as the core language of the Japanese 'fifth generation' computers. It attempts to describe and solve problems in logical terms that are not machine-dependent. It also has potential for parallel processing.

PROM Programmable (or Programmed) Read-Only Memory. Any type of memory device that is not recorded during manufacture, but requires some form of physical operation to be programmed. Special processes can allow some PROMs to be erased and reprogrammed (*see* UV PROM). They are used to form read-mostly memories (RMM) of which the most common type is the floating-gate PROM, which employs the metal-oxide semiconductor (MOS) as its basic memory cell. The cell has two gate electrodes, one above the other, separated by a layer of silicon dioxide. The lower gate is completely surrounded and therefore is floating.

PRR Pulse Repetition Rate. The number of pulses that occur in a unit time experienced at a particular point in a transmission medium of free space, or at a point in a circuit (CCT), device or system. The rate is normally expressed in pulses per second (pps,1).

PRT (1) Production Run Tape. A tape containing a verified and checked schedule for a production run, and able to run on a variety of computers.

(2) Program Reference Table. A particular memory-storage area, used to store operands, array references, file references and references to segments of a computer program. It allows programs to be independent of the memory locations actually occupied by parts of the program and data.

ps PicoSecond. One millionth of a second (s), that is 0.000 000 000 001 or 10^{-12} second.

PSD (1) Post Sending Delay. A time interval that occurs between the finish of dialling on a telephone and the reception of a return signal, which could be a ringing tone or a busy tone (BT,2).

(2) Professional Development System. PDS has been launched by the British Computer Society (BCS) and supported by the Manpower Services Commission. It provides a framework by which those either starting or established in the Information Technology (IT) field can progress to a 'Professional Status' in a structured, logical and validated way.

PSE Packet-Switching Exchange. A group of network nodes (intersections of a number of networks), at one location. The splitting of information into packets for transmission is often done at the PSE. *See* PSS,1, packet-switching system.

PSK Phase-Shift Keying. A process whereby the phase of a modulated carrier wave (CW,1) is changed between a specified set of predetermined values.

PSN Public Switched Network. A switching system made available by a common carrier for circuit switching many subscribers. An example in the UK is British Telecom (BT,1).

PSS (1) Packet-Switching System. A data transmission system that provides for the transfer of discrete groups or packets of information along a channel (CHNL(a)), which is occupied only for the duration of the transmission of the packet. The packet itself contains address information and the channel is available for the transfer of other packets, providing communication through the network to other users without the need for a two-way transmission path.

(2) Packet-Switched Service. The packet-switched network commissioned in 1980, and run by the Post Office (PO) in the UK.

(3) Packet-Switch Stream. *See* PSS,1, packet-switching system.

PST Pair-Select Ternary. A pseudo-ternary code where pairs of binary digits (BITs) are coded together, making a third value, in such a way that the resultant signal has no long strings of zeros. A ternary code has three possible values, for example +1 volt (V,1), O and −1 V.

PSTN Public Switched Telephone Network. *See* PSN.

PSU (1) Power Supply Unit. All electronic devices and circuits require some form of external power source. The voltages required range from 1.5 volts (V,1) to 25 kV or higher, and the currents can range from a few microamps (mA) to several tens of amps (A). PSUs can be produced to cover all these requirements. Most electronic circuits run off the 240 volts, 50 hertz (Hz) supply via a PSU, the simplest form of which consists of a transformer that produces alternating current (AC,1) at a lower voltage (V,2), which is in turn rectified by a rectifier circuit to give off direct current (DC). The high ripple content of the rectifier's output (O,2) is removed by a smoothing circuit to provide a stable supply. Also some form of stabilization is provided as the DC output has to remain substantially constant free from the effects of changes in load current, mains input and temperature. A simplified schematic diagram is shown in *Figure P.5*.

(2) Program Storage Unit. A chip (CHIP) that principally provides for the storage of programmed instructions and various nonvolatile data constants of a program. It is able to be

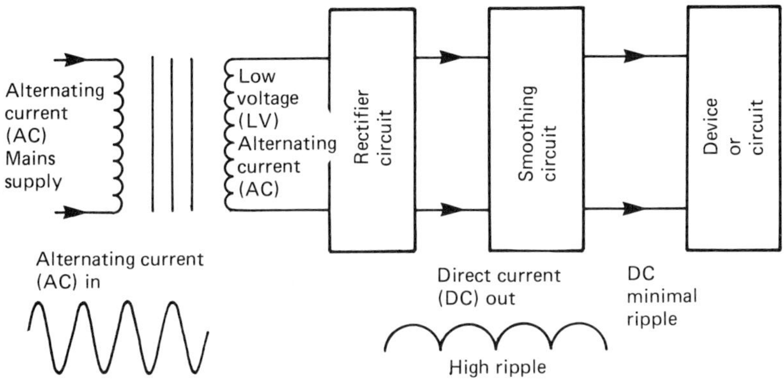

Figure P.5 A simplified schematic diagram of a power-supply unit (PSU)

interfaced directly with a central processor unit (CPU) without the need of buffer circuits.

PSW (1) Processor Status Word. A register held in a central processor unit (CPU) containing the status of the most recent instruction execution result. It can record error and exception conditions that are able to be dealt with by the operating system (OS).
(2) Program Status Word. A sector of main memory storage that is used as a computer system status indicator in relationship to a specific program.

Pt PlaTinum. An extremely stable noncorroding metal with the atomic number 78. It is used as a conductor or contact at high temperatures or where chemical attack is expected.

PT Paper Tape. A strip of paper normally 25 mm wide and up to 300 m in length on which data is recorded in the form of holes arranged in groups of binary format along the tape's length. The tape is divided into columns which run the length of the tape and channels which extend across the tape width, the number of which depends on the system used.

PTC Positive Temperature Coefficient. The attribute of a semiconductor material to have a resistance that increases with an increase in temperature. PTC semiconductor materials are normally employed in thermally sensitive resistors (THERMISTOR).

PTFE PolyTetraFluoroEthylene. A thermoplastic that can be used over a wide range of temperature and is very resistant to moisture. It has excellent electrical insulating properties and can be moulded or shaped into many forms that are used in the electrical industry.

PTM Pulse Time Modulation. A term, used generally, to

describe various methods of modulation in which a pulse carrier that is time-related varies according to the instantaneous value of a modulated signal.

PTP (1) Point-To-Point. The transmission of data between two points. It can also relate to the single physical line, switched or non-switched, that connects a remote terminal to a computer. (2) Paper-Tape Punch. An output (O,2) device that punches or encodes machine-readable output data in the form of holes into paper tape (PT). It has been a very popular device for producing physical data storage. (*See* PTR.)

PTR Paper-Tape Reader. A popular low cost input device that senses and translates the holes or information provided in a paper tape (PT) into machine code. The use of a paper-tape punch (PTP) and a PTR combination has now been generally superseded by magnetic tape (MAGTAPE) cassette units.

PTS Program Test System. Capable of automatically verifying and checking programs, this specific system is able to assist in the organization of a production run by providing diagnostic information.

PTT (1) Postal, Telegraph and Telephone authority. The particular national authority that is in overall command of posts and telecommunications within a particular country. In the UK the General Post Office was the PTT, until it was replaced by two authorities, the Post Office (PO) and British Telecom (BT,1). In the USA there is no corresponding body, although the Federal Communications Commission (FCC) regulates these matters.
(2) Push-To-Talk. A switch that is depressed during transmission and released during reception, normally on the handsets of two way radio communication devices such as citizen band radios (CB), ship-to-shore radios and police communicators. Also known as 'pressed switch'.

p-type An extrinsic semiconductor containing a higher density of mobile holes (Valence band electrons) than of conduction electrons, which means the holes are the majority carriers, *see Figure P.6*. Compare n-type semiconductor.

PU (1) Per-Unit. In the per-unit system quantities are expressed in decimal fractions of unity (one) rather than as parts per 100 (per cent). Advantage is gained in that quantities expressed in per-unit values can be more easily multiplied and divided, making calculation simpler.
(2) PickUp. In audio and video, a transducer for extracting audio frequency (AF) or video signals from disc recordings.

PUFF A colloquial expression used for picofarad, it is derived from the markings of 'pF' on capacitors representing picofarads.

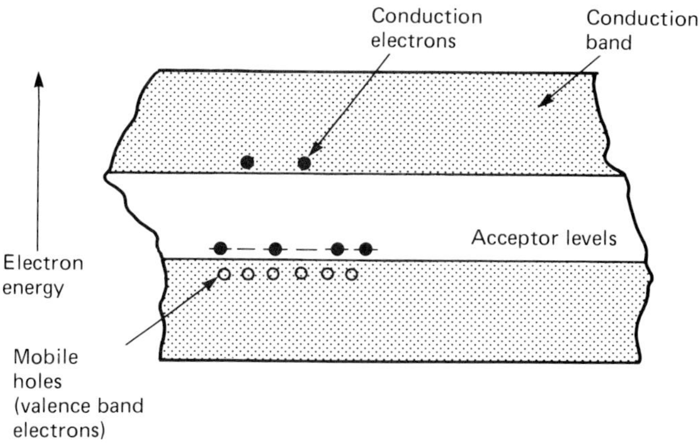

Figure P.6 Energy bands of a p-type semiconductor

PUFFT Purdue University Fast ForTran. A particular version of the formula translator (FORTRAN) computer language.

PUJT Programmable UniJunction Transistor. A programmable version of the unijunction transistor (UJT).

PUSH An instruction that is used to deposit a word on the top of a stack or push-down list (as opposed to POP).

P-V Peak-to-Valley value. The comparison or difference between the greatest instantaneous value (peak value) and the least instantaneous value (valley value) of a time-dependent quantity.

PVC PolyVinylChloride. A thermoplastic that is available in a wide range of bright colours and can be transparent or opaque. Being resistant to water, corrosion and mechanical abrasion, it is used widely in the electrical industry for sheeting and insulation of extruded cable.

PVDF PolyVinylDene Fluoride. A semicrystalline, high molecular weight polymer of the repeat unit, CM_2-CF_2, that is used as transducer film material. It is approximately 50% crystalline and 50% amorphous and has a high piezo-electric response.

PVT Page View Terminal. A visual display unit (VDU) that allows the user to display (D,6) pages of text, one page at a time.

PW Private Wire. A specific channel (CHNL(d)) along with the necessary equipment, that is intended for the exclusive use of a particular subscriber. Application of such a line can include a telephone link, a facsimile (FAX) or telecommunications link.

PWB Printed Wire Board. A conductive pattern that has been formed, by etching or plating, on the surface of an insulating baseboard. *See* PCB, printed circuit board.

P-wire Private WIRE. A wire over which automatic switching functions of guarding, holding and releasing are controlled and also in some systems over which metering may be achieved.

PWM Pulse Width Modulation. Pulse modulation (PM,2) in which the duration of each pulse in a pulse train is varied according to the instantaneous value of the modulating wave. The modulating signal can vary when the leading edge, the trailing edge or both edges of each pulse occurs.

PWR (1) PoWeR. The rate at which energy is absorbed or radiated by a subassembly or device. Unit is the watt (W,2). *See also* P,3.

(2) Pressurized Water Reactor. A particular type of nuclear reactor that uses pressurized water as its coolant.

PX Private eXchange. A telephone exchange that serves a particular organization and has no means of connection with the public telephone network.

PZT PieZoelectric Transducer. A crystal that produces a voltage (V,2) when stressed by compression, expansion or torsion. The voltage is only produced when the applied stress is changing.

Q

Q (1) Queue. Arranged in sequence, a collection of items that has two ends which are the head and the tail. New items are added to the tail. Items can be removed from either the head or the tail.

(2) A register used as an accumulator (ACC(b)) extension. This is necessary for efficient multiple-divide computer programming.

(3) An address, or source location in internal storage in some forms of computer equipment from which data can be transferred.

(4) A comparison test made between two or more units of quantitative data to discover if they are equal or unequal.

See also Q-axis, Q band, Q-component, Q-factor, Q-meter, Q-signal, Q-switch.

QA Quality Assurance. Methods employed in a systematic manner to provide the necessary confidence that an item itself has been tested or is constructed of components which have been tested and will perform satisfactorily.

QAGC Quite Automatic Gain Control. A delayed automatic gain control (AGC) combined with a signal suppressor in such a way that the output (O,2) of a receiver is suppressed for inputs (I/P) too weak to operate the AGC.

QAM (1) Quadrature Amplitude Modulation. A high-speed modulator/demodulator (MODEM) modulation technique used in digital (D,4) transmission where two sinusoidal carrier waves having a phase difference of 90° between them, are amplitude modulated (AM) by independent signals and the resulting components are summed to produce the output (O,2) signals. This system is used in colour television (TV(a)) transmission.

(2) Queued Access Method. A term applied to any data access method that is able to synchronize the data transfer between the user program and input/output (I/O) devices and reduce input/output delays.

Q-axis The chrominance axis that corresponds to the coarse chrominance primary in the National Television System Committee (NTSC) colour television transmission (TV(a)) system. Also known as narrow-band axis.

Q band Microwave frequencies. Q-band microwave (μW) frequencies occur between 36 and 46 gigahertz (GHz) and form part of a larger continuous range of frequencies covering 0.225 to 100 gigahertz which are usually subdivided into bands designated by letters. Q-band frequencies have wavelengths of between 0.834 and 0.652 cm.

QC Quality Control. Workmanship, processes and materials that have been monitored by an organized and formal process to produce consistent and uniform products. The process can involve the keeping of evaluation records, batch logs, and error summary reports.

QCB Queue Control Block. A memory storage location that holds information in a condensed form, which is used to regulate the sequence in which entries in a list waiting for service are processed.

Q-component Resulting from the modulation of a sub-carrier by the Q-signal, this is the component of the chrominance signal in the National Television System Committee (NTSC) colour television broadcasting system.

QED Quick text EDitor. A high-speed system that provides the user with a flexible source text generation system. Source statements are entered which can be output (O,2), statements added, or groups of statements altered.

Q-factor Quality-FACTOR. A measure of the ratio of energy stored in a system, circuit, component or device in comparison to the energy dissipated. The energy dissipation could be due to resistance (R,1), electromagnetic radiation or other factors.

QI Quartz Iodide. Refers to a type of lighting system.

QIL Quad-In-Line. A dual-in-line package (DIP) with four similar functional circuits or units (two rows of pins on each side). *See* QUIP.

Q-meter An instrument used to measure the Q-factor (*see* Q-factor) of a tuned circuit or of circuit components.

QPDM Quad Pixel Dataflow Manager. A processor that integrates text and graphics in bit-mapped graphics systems, supporting display (D,6) screens. The QPDM performs three key functions demanded by any graphics system: refreshing the display, updating the display memory, and refreshing the dynamic random-access memories (DRAM) in the display memory.

Q-point Quiescent POINT. The area on the characteristic curve of an active device, such as a transistor, when the device has no input signal.

QPL Qualified Products List. A list used in the USA containing products that are considered to be of high reliability and are qualified for military (MIL,2) use.

QPP Quiescent Push-Pull. A circuit that operates in a push-pull mode using two pentodes biased near anode-current cut off. The circuit was once often used in battery-operated equipment.

QSAM Queued Sequential Access Method. A type of the basic sequential access method (BSAM), which uses blocks formed from queues (Q,1) of input data which are waiting for processing, or output (O,2) data blocks which have been processed and are waiting for transfer to output devices or other auxiliary storage.

Q-signal The signal that represents the chrominance information along the green-magenta axis in the National Television System Committee (NTSC) colour television (TV(a)) transmission system.

Q-switch A device that prevents pulsed laser (LASER) emission being achieved until a certain required energy level is reached in the lasing material. It provides shorter and more intense pulses at a higher repetition rate than could be reached by pulsing the lasing material.

QTAM Queued Telecommunications Access Method. A method of transferring data between main computer storage and remote terminals. It uses an application program with a specific macro instruction that requests the transfer of data, which is performed by a message control program that also synchronizes the transfer, which in turn eliminates delays for input/output (I/O) operations. The technique is employed for data collection, message switching and other teleprocessing uses. It provides the capabilities of the basic telecommunications access method (BTAM) as well as the ability of queued messages on

Figure Q.1 A QWERTY keyboard (*courtesy* Alphameric Keyboards Ltd)

Figure Q.2 A QWERTZ keyboard (*courtesy* Alphameric Keyboards Ltd)

direct-access storage devices.

QUICKTRAN QUICK forTRAN. A development of the formula translation (FORTRAN) programming language designed primarily for running from a remote terminal.

QUIP QUad-In-line-Package. Having two rows of pins along each longitudinal edge of its package, rather than one row as in a dual-line package (DIP), this 64-lead integrated circuit (IC) package was developed jointly by Integrated Electronics (INTEL) and 3M.

QWERTY A standard (English-language) computer terminal keyboard (KB,1) or typewriter keyboard on which the keys are arranged such that the top row below the numerals begins with the letters Q, W, E, R, T and Y, reading left to right, as shown in *Figure Q.1.* Mainly used in the UK and USA as well as other English-speaking countries. (*See* AZERTY and QWERTZ.)

QWERTZ A German-language computer or typewriter keyboard (KB,1) with the keys arranged so that the top row below the numerals begins with the letters Q, W, E, R, T and Z, reading left to right, as shown in *Figure Q.2.* It normally contains some additional letters peculiar to the German language, as well as instructional keys notated in German. It is used mainly in German-speaking countries. (*See* AZERTY and QWERTY.)

R

R (1) Resistance. Having the unit ohm, this is a measure of the tendency of a particular material to resist the passage of an electric current (I) and to dissipate electrical energy into heat energy. It is the ratio of applied potential difference (PD,1) across a conductor to the current flowing through it.
(2) Resistor. The symbol R is used to represent resistor on parts lists and to annotate schematic diagrams.
(3) Reluctance. Magnetic resistance, the ratio of magnetomotive force to the total magnetic flux having the unit henry $^{-1}$.
(4) Rontgen. A measurement of the effect of ionization. It can be defined as the quantity of gamma- or X-rays such that associated emissions every 1.293 mg of air produce ions in the air that carry 1 electrostatic unit of charge.

RACE Random-Access Computer Equipment. Computer equipment in which information can be introduced and stored, where any location can be read from, or written into, in a random access method. It is used where the information to be obtained is unrelated to the location of previous information.

RAD (1) Rapid- (or Random-) Access Disk. A memory storage device, often a magnetic disk that provides rapid access to a memory location, and retrieval of data, a procedure that is often synonymous with random access or direct access. It is contrasted with sequential access in that it does not depend upon the access of preceding data.

(2) Radiation Absorbed Dose. The base unit of the amount of radiation absorbed which results in the ionization of the material onto which it falls or strikes the surface of. A dose of 1 RAD equals the absorption of 100 ergs of radiation energy per gram (g) of absorbed material.

R&D Research AND Development. The application of scientific and technical activities directed towards the development of new products systems or services.

RADA Random-Access Discrete Address. A particular location in a random access memory (RAM).

RADAR RAdio Direction And Ranging. Using reflected radio waves of microwave (μW) frequency, a RADAR system is able to locate distant objects. Current sophisticated radar systems can provide information concerning both stationary and moving objects and can be used for navigation and guidance of ships, aircraft and other vehicles and systems.

RADIR Random-Access Document Indexing and Retrieval. A system for creating an index of documents, in which access to data to be obtained is unrelated to the location of previously accessed information.

RALU Register and Arithmetic Logic Unit (or Register-equipped Arithmetic Logic Unit). An assembly of logic elements including accumulators (ACC(b)), stack and arithmetic logic units (ALU) which together provide both storage and processing functions. All the elements of the unit are married by a set of buses.

RAM Random-Access Memory. A direct-access memory device that is used where it is required to change the binary digit (BIT) location during the routine operation of the memory. It is a mass storage device designed so that the location of stored data is independent of content and any location may be directly accessed without having to work through from the beginning. The fast access to any storage location is provided by means of vertical and horizontal co-ordinates and information is written in or out using the same procedure. Sometimes also known as a 'random-access/read-write memory'.

RAMIS Random-Access Management Information System. A system providing information in response to users' requests in which the computer files store management data in a random-access memory (RAM).

RAMTOP Random-Access Memory Top. The highest address at which program data can be held in the random-access memory (RAM) of a microcomputer. Also known as HIRAM.

RAS Row Address Strobe. Used in dynamic random-access memories (DRAM) this signal is employed to reduce the number of pins necessary, by multiplexing the address.

RATFOR RATional FORtran. A structured development of formula translator (FORTRAN) which is translated or compiled into standard FORTRAN by a processor.

RAYNET Radio Amateurs emergencY NETwork. An organization set up to provide assistance to the authorities during disaster relief operations, by providing a means of communication.

RAX (1) Remote ACCess. The use of a terminal that is capable of providing access to a computer from a point that is physically distant to the computer.
(2) Rural Automatic eXchange. An automatic telephone exchange located in the countryside. Due to its position it deals with a lower average number of telephone calls than an automatic exchange located in a town or city.

RB Return to Bias. A particular mode of recording where the state of the medium used for recording alters from a bias state to another state, and then returns to record a binary 1.

RBE Remote Batch Entry. The transfer of a group of records to be processed as a single item, over a distance by means of a communications link.

RBSOA Reverse Bias Safe Operating Area. Normally refers to the bias drive circuits employed in power (mains) switching.

R-C Resistance-Capacitance. A prefix used to describe devices or circuits where the use of resistance (R,1) and capacitance (C,5) is necessary for their operation, or that employ resistance-capacitance coupling.

RC Regional Centre. Class 1 of five class numbers assigned to offices according to their function in the American direct distance dialling (DDD) telephone network. Any one centre can handle telephone call traffic from one, two or more centres lower in the hierarchy, Class 1 being the top classification.

RCCB Residual Current Circuit-Breaker. A term used in switching terminology that is superseding earth leakage contact-breaker (ELCB) and conforms to current regulations. RCCBs can provide users with protection against serious electrocution either from appliance or cable by tripping out well before a serious shock can be received, typically in 30 milliseconds and at 30 milliamps.

RCTL Resistor-Capacitor-Transistor Logic. The logic elements of a circuit or system which are similar to those of a resistor-transistor logic (RTL) circuit, except that capacitors (C,6) are provided to enhance switching speeds.

RCV ReCeiVe. The process of obtaining a transmitted signal, as contrasted with sending or transmission.

RCWV Rated Continuous Working Voltage. The maximum (MAX) voltage (V,2) that can safely be applied to a resistor. For example, the maximum RCWV for a 50 ohms 2 watts (W,2) resistor is 10 volts (V,1).

RD ReaDer only. The function of a paper-tape reader (PTR), which is a device that senses and translates the holes or information on a paper tape (PT) into computer machine code, but is unable to punch holes or put information on blank tape.

RDF Repeater Distribution Frame. Housed in a telecommunication repeater station, this frame provides interconnection for various items of apparatus or equipment which include signalling units, amplifiers (AMP) and transformers (XFMR).

RDOS Real-time Disk Operating System. A disk operating system (DOS) that operates in real time (RT,1).

RDY ReaDy. The control signal, used in some systems, for slow memory to indicate if data is valid.

READ (1) Real-time Electronic Access and Display. The acquisition of information from a storage device or a peripheral unit, which is produced on a visual display unit (VDU). The whole operation takes place in real time (RT,1).
(2) Remote Electrical Alphanumeric Display. A visual display, normally on a visual display unit (VDU), of numerical and alphabetical details, that is provided at a location physically distant from the computer.

READLN READ LiNe. A statement in the Pascal (PASCAL,1) computer programming language that differs from 'READ' in that after READLN is executed the computer will then be waiting at the beginning of a new line for any further data input (I/P).

REALCOM REAL-time COMmunications. A communication system in which response occurs at the same rate as that of the data input (I/P).

RECOL REtrieval COmmand Language. A computer language employed to interrogate the store of data held on the computer files within an information retrieval system (IRS).

REED Restricted Edge Emitting Diode. A light-emitting diode (LED) that emits light over only a small area of one edge. This region of restricted light emission, when used with integrated optical circuits (IOC,2) and optical fibres (OF), improves coupling efficiency.

REGEN REGENerative repeater. A device used to amplify, retime and retransmit received signal impulses at their original strength or level.

RELP Residual Excited Linear Predictive coding. An algorithm used with a coder-decoder (CODEC) developed for use in digital (D,4) cellular radio devices.

REM (1) REcognition Memory. A simplified associative memory designed to interface simply with ordinary computers.
(2) REMark. A statement in the beginners' all-purpose symbolic instruction code (BASIC) computer programming language that is used to include a remark or comment, for example to put in a program title. It is for convenience only, being ignored during the execution of the program.

REMSTAR Remote Electronic Microfilm in Storage Transmission And Retrieval. A computer output microfilm (COM) station that is physically distant from the main system. It is able to provide necessary distribution and storage functions.

RENUM RENUMber. A command in the common algorithmic language (COMAL) computer programming language which renumbers the lines in a program.

RES REStore. (a) To return a device or circuit to its initial state. (b) To return a counter or timer to zero. (c) To return a computer word to its initial or preselected value. (d) To return a register, trigger or error correction signal to zero or to a specific initial condition. (e) To place a zero in a binary cell.

RETD Red Especial de Transmission de Datos. The Spanish postal telegraph and telephone authority (PTT) packet-switching network.

REW REWind. To return a magnetic tape (MAGTAPE), film or paper tape (PT) to its beginning, or some other initial data point.

REX Real-time EXecutive. A sequence of operations that a computer is to perform, in which the computer responds at the same rate as that of the data input (I/P).

RF Radio Frequency. A frequency of electromagnetic radiation or alternating current (AC,1) within the range of 3 kilohertz (kHz) to 300 gigahertz (GHz). By international agreement radio frequencies occupy the bands numbered between 5 and 11 (inclusive) and have wave lengths 100 km and 1 mm. Each band extends from 0.3×10^N hertz (Hz) to 3×10^N Hz where N is the specific band number.

RFC Request For Connection. Required to achieve a connection, this command is used in the formalized method by which two parties communicate in an advanced research projects agency network (ARPANET) packet-switched computer network.

RFI Radio Frequency Interference. The distortion or impairment of a desired transmission or signal brought about by radio frequency (RF) disturbances such as atmospheric noise.

RG Release Guard. The protection of a piece of telephone equipment from seizure (i.e. calls to be established) or attempted use whilst it is returning a line from a busy state to a free state.

RGB Red, Green, Blue signals employed in colour television (TV(a)) transmission, these three different signals each represent one of the three primary colours. Also known as 'tristimulus' signals.

RGS Release Guard Signal. This signal protects a telephone circuit or channel (CHNL(f)) from seizure (i.e. calls being established) at its outgoing end whilst equipment at its incoming end is returning from a busy to a free state.

RI Radio Interference. A phenomenon experienced when an unwanted radio signal or radio frequency (RF) disturbance such as 'noise' or 'atmospherics' interferes with a wanted or required radio signal.

RIAA Record Industries Association of America. An organization in the US concerned with the standards applying to the production of records and the record industry, such as the recording characteristics of discs.

RICASIP Research Information Centre and Advisory Service on Information Processing. An organization in the USA sponsored by the National Science Foundation and the National Bureau of Standards (NBS). Its aims are to provide aid and assistance concerning information processing.

RIG A slang expression used to describe transmitter or transmitter receiver (TRANSCEIVER) equipment.

RISC Reduced Instruction Set Computer. A concept that helps achieve high processor performance.

RIT Receiver Incremental Tuning. A control that permits the frequency received in a transmitter/receiver (TRANSCEIVER) to be carried (typically ±5 kilohertz (kHz)) on one side or the other of the transmitted frequency.

RITA Rand Intelligent Terminal Agent. A RITA system permits a terminal simple access to a computer network.

RJE Remote Job Entry. The action of a remote device providing input of job information to a main computer system.

RLD ReLocation Dictionary. A section of a program that contains the information necessary to alter the label identifying the location of where it is stored, after it has been placed in a new location, i.e. relocated.

RM Regular Matrix. A means of encoding four audio channels

(CHNL(g)) into two and then decoding into four. It is a system used by some Japanese manufacturers.

RMM Read-Mostly Memory. A memory device used where long-term storage of information is required. Thus, stored information is altered very infrequently, so allowing for a device that has the write time much greater than the access time. (*See* ROM.)

RMS (1) Recovery Management Support. When operations fail due to some hardware error, this facility gathers information concerning hardware reliability and allows the operations to be attempted again.
(2) Root-Mean-Square. The square root of the average of the square of a variable. It is usually employed to express the effective value of an alternating current (AC,1). The RMS value of an alternating current is the square root of the average of the square of the current taken throughout one period. This is comparable to a direct current (DC,1) in terms of heating effect, etc.

RMW Read-Modify-Write. A refresh cycle used in the operation of a random-access memory (RAM).

R/NAV Radio NAVigation system. A computerized avionic system for small military and commercial aircraft, designed to work with conventional cockpit instruments. It enables a pilot to fly along any course without manual calculations or reference to maps and is able to provide dead reckoning navigation when temporarily outside the range of a radio beacon.

RND RaNDom. An arithmetical function in beginners' all-purpose symbolic instruction code (BASIC), it generates random numbers between limits. RND(1) would generate random numbers between 0 and 0.999.

RO (1) Read Only. Data access that allows the data to be examined or read, but does not permit any modifications to be made.
(2) Receive Only. A printer unit in data processing (DP,1) or a radio receiver in telecommunications (TELECOMMS) that can only receive messages and is unable to generate them.

ROI Return on Investment. The ratio of the amount of money one invests to the amount of money one receives in revenue.

ROM Read-Only Memory. A computer memory with a permanent set of information written into it, whose contents under normal conditions are not intended to be alterable by instructions. There are three groups of ROMs: masked programmed, fusible link (FL) and alterable. The mask programmed ROM is programmed by a mask pattern, as a part of its final manufacturing process. The fusible link is permanently programmed by electrical pulse or mechanical means. The alterable device undergoes changes which are induced electrically.

ROOST Rapid Optical Ocean Surveillance Testbed. A research and development (R&D) programme to refine optical signal processing techniques for use in ocean surveillance, sponsored by the US Navy and the US Defense Advanced Research Projects Agency (DARPA). It is a hardware-orientated engineering programme to test the concept that successful integration of advanced optical, digital and display technology can increase the speed with which acoustic data can be obtained in ocean environments.

ROPP Receive-Only Page Printer. Used at communications stations that are unable to generate messages, this type of teleprinter unit has a printer only.

RP (1) Reader Punch. The function of a paper tape (PT) reader punch. It can punch holes in blank paper tapes and is also capable of sensing and translating holes on a tape into computer machine code.
(2) Reduced Propogation. Refers to cables insulated and sheathed with materials which reduce the tendency for fire to propogate along the length of the cable.

RPG Report Program Generator. (a) A processing program that provides a convenient method for producing a wide variety of reports. The generator can vary from a listing of a magnetic tape (MAGTAPE) reel to a very formal, calculated and edited table of data provided from a number of inputs. (b) A computer programming language that helps the user by providing facilities for assimilating data, producing printed reports and creating, updating and maintaining files. Also known as RePort Generation.

RPM Revolutions Per Minute. The total number of, or fraction of, complete turns or cycles made by a rotating device in a minute.

RPN Reverse Polish Notation. A form of logic that allows the user to enter every problem from left to right as it is written or notated. The RPN calculator has no *equals* key, as it uses an *enter* key during operation. For example, to calculate 3 multiplied by 7 the RPN calculator functions thus:

3 PRESS	ENTER PRESS	× PRESS	7 PRESS	ENTER PRESS	21 DISPLAY

on pressing 'enter' the second time the calculation takes place and the answer is displayed.

RPS Revolutions Per Second. The total number of, or fraction of, complete turns or cycles made by a rotating device in a second.

RRRV Rate of Rise of Restriking Voltage. The speed at which the voltage (V,2) increases across the contacts of an interrupting device. Across an arc initially a very small current flows due to residual ionization. If the RRRV is sufficiently fast the residual ionization increases, and the current is re-established. On the other hand, if the RRRV is too slow, the ionization decays and the current cannot re-establish itself.

RS (1) Record Separator. A particular control character that is used to indicate logical divisions between separate areas of records.

(2) A simple form of flip-flop having two inputs, one 'set' and the other 'reset'. It consists of two cross-coupled not and (NAND) or not or (NOR) gates and stores a binary digit (BIT) controlled by input signals. (*See* SR.)

RS-232C A popular and widely used standard for the interconnection of computer terminals, cathode-ray tubes (CRTs), displays (D,6), modulator/demodulators (MODEMs), teletypewriters (TTYs), input/output (I/O) devices, peripheral devices, data terminal equipment (DTE) and data communication equipment (DCE). RS-232C is actually an electrical standard and the suffix C denotes the latest revision.

RS-449 An interface standard. It enhances the capabilities defined in the popular and widely used RS-232C (RS-232C).

RSGB Radio Society of Great Britain. The UK national society for radio amateurs which endeavours to promote greater awareness, understanding and use of amateur radio.

RT (1) Real Time. A computer system response which occurs at the same rate as that of the data input. Remote controls or automatic guidance systems by necessity operate in real time.

(2) Radio Telephone. A combined and complete radio receiver and transmitter (TRANSCEIVER) required at one station for two-way communication.

(3) Remote Terminal. A terminal that is physically distant from the process or computer system it is accessing. The link can be made via a cable or a broadcast transmission.

(4) Reperforator/Transmitter. A teletypewriter (TTY) that consists of a reperforator and a tape transmitter which are independent of one another. It performs a relaying function and it can transform the speed of incoming information to a different outgoing speed, and is also used for temporary queueing (Q,1).

(5) RighT. A command in the Logo (LOGO) computer

programming language, to move a Turtle graphics robot right through a given number of degrees.

RTBM Real-Time Bit Mapping. A process employed to aid production. It can be used to locate defects in the masks used for the diffusion, oxidization and metallization steps in chip (CHIP) manufacture, and can also be used to overcome process problems.

RTC Real-Time Clock. A device that produces readable digits (DGTs) or periodic signals that allow a computer to recognize elapsed time between events, and also indicates the performance of time initiated or time related processing.

RTE Real-Time Executive. A system of computer software that provides an overall guidance for scheduling, handling and program capabilities.

RTI Radiation Transfer Index. An index given as the 1/10th root of the overall efficiency of a radiation transmission system.

RTL Resistor-Transistor Logic. The logic elements of a circuit or system which comprise resistors (R,2) and transistors, as shown in *Figure R.1*. There is a family of RTL integrated logic circuits

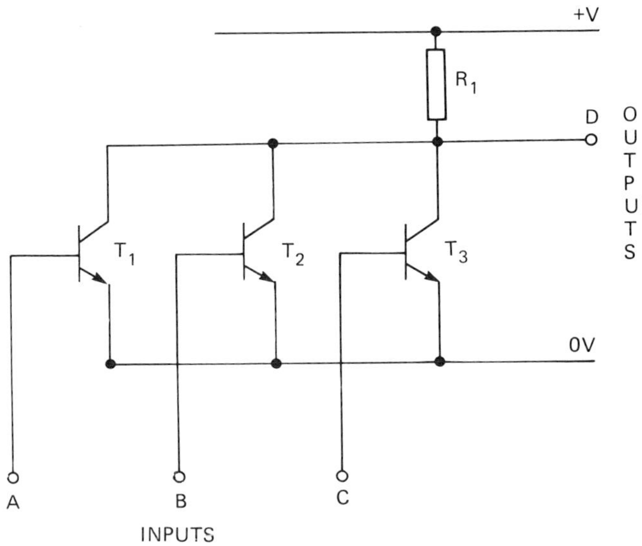

Figure R.1 A resistor-transistor logic (RTL) gate

in which the input is through a resistor into the base of an inverting transistor. The basic Not OR (NOR) gate is one. RTLs were the first integrated logic family to be developed but are now little used.

RTM Register Transfer Module. A functional register unit designed for the construction of register logic systems.

RTOS Real-Time Operating System. Software that is required to manage the hardware resources of a computer system, which is capable of real time (RT) task management. The management can include scheduling, real-time event counters and event scheduling.

RTT Regie des Telegraphes et des Telephones. The postal, telegraph and telephone (PTT) authority of Belgium.

RTTY Radio TeleTYpewriter. A teletypewriter (TTY) that operates via radio circuits employing frequency-shift keying (FSK) and also audio frequency-shift keying (AFSK).

RTU Remote Terminal Unit. A computer terminal that is physically distant from the main system that it is accessing. The link can be made via a cable or a broadcast transmission.

RTV Room-Temperature Vulcanizing elastomer. A covering, normally made from a silicone rubber compound, that is applied over an entire chip (CHIP) to protect it from its surroundings.

RUBOUT A colloquial expression for 'delete character' (DEL,2).

R/W Read/Write. The function of the small electromagnetic heads which read, record (write) or erase polarized areas on a magnetic surface, usually tape.

RWD ReWinD. A computer program term that is intended as an instruction to rewind a reel of tape.

RWG Ridge WaveGuide. Similar to the inverted rib guide (IRW) laser (*see* LASER). The configuration of the RWG achieves the low threshold characteristics of buried heterostructure (BH), but with lower production costs.

R-wire Ring WIRE. A wire in a telephone exchange circuit that is connected to the ring-shaped contact of a telephone jack plug or with the corresponding contact of a switchboard jack.

RWM Read/Write Memory. Where the application for a small computer system is fixed, it would have a read-only memory (ROM) program memory. Where it is necessary to change the system application by an operator a RWM is employed. The RWM has provided advancement in small computer system complexity for hardware and programs.

RX Receiver. A unit or circuit in a communications system that converts incoming electrical signals into audible signals. Normally on communications equipment a green panel indicator shows that an RX circuit is in operation.

RZ Return-to-Zero. A signal presented in a digital (D,4) form having pulses which are not full symbol length, i.e. they are shorter than the spacings between the symbols, as shown in *Figure R.2*. Intervals occur between successive binary 1 symbols, i.e. the signal returns-to-zero.

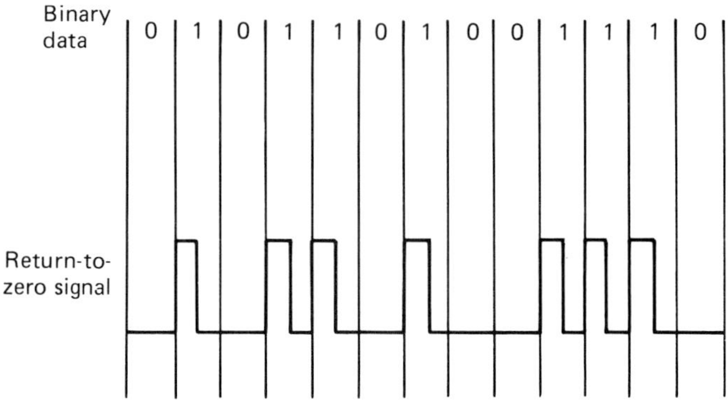

Figure R.2 A return-to-zero (RZ) signal

S

s (1) Second. The unit of time in the Système International d'Unités (SI) system of units, it is defined as the duration of 9 192 631 770 periods of the radiation of the cesium – 133 atom. Formerly it was defined as 1/86 400th of the average solar day. (2) Source of a field-effect transistor (FET). The connection to the channel (CHNL(h)) from which the majority carriers enter the channel.

S Siemens. The unit of electrical conductance in the Système International d'Unités (SI) system of units, susceptance and admittance. A substance possesses a conductance of one siemen if its electrical resistance (R,1) is one ohm.

S100 An interconnection system for computers which consists of 100 pins and has been designed for the Intel 8080 and similar microprocessors.

SALINET SAtellite Library Information NETwork. A satellite communications system used to provide library services to the remote areas of Canada.

SAM (1) Subsequent Address Message. A message used in common-channel signalling (CCS) following an initial address. It contains one or more additional digits (DGTs) of address information along with or instead of an end-of-pulsing signal. (2) Serial-Access Memory. A memory-storage device in which the information is entered in sequence. This results in sequential processing of the data.

SAMICS Solar Array Manufacturing Industry Costing Standards. A technique that employs a computer program, providing a common ground for the comparison of the economic aspects of several alternative approaches to the production of silicon cells. It can provide some of the best documented and validated costings of the process sequences involved in fabricating solar modules.

SAT SATurated mode. The characteristics of operation of a bipolar or field-effect transistor (FET) beyond its pinch-off voltage, i.e. where the pinch-off voltage is less than or equal to the drain voltage. The drain current is independent of the drain voltage in the region.

SAW Surface Acoustic Wave. Waves generated on the surface of piezo-electric solids along which they also propagate. They have a velocity that is independent of frequency, but comparable to the velocity of sound through the material. SAW devices are used in bandpass filters.

S band Microwave (μW) frequencies that occur between 1.55 and 5.20 gigahertz (GHz) and form part of a larger continuous range of frequencies covering 0.225 to 100 gigahertz (GHz) which are usually subdivided into bands designated by letters. S-band microwave frequencies have wavelengths of between 19.3 and 5.77 cm.

SBC Single-Board Computer. A complete computer that is produced on a single printed circuit board (PCB), *see Figure S.1*, including a central processor unit (CPU), random-access memory (RAM) and all necessary interfaces. It is widely used by original equipment manufacturers (OEM) where an element of intelligent control is required to be built into equipment.

SBR Styrene-Butadiene Rubber. A material used for cable insulation.

SBS Satellite Business Systems. An organization established to offer satellite communication channels (CHNL(a)) to provide the USA with an electronic mail service.

s/c Short Circuit. *See* SHORT.

SC Sectional Centre. Class 2 of five class numbers assigned to offices according to their function in the American direct distance dialling (DDD) telephone network. Any one centre can handle telephone call traffic from one, two or more centres lower in the hierarchy, Class 1 being the top classification.

SCA Steel-Cored Aluminium. Consisting of a core of galvanized steel strands encircled by layers of aluminium, this conductive material combines the strength of steel with the conductive ability of aluminium.

SCARAB Submersible Craft Assisting Recovery And Burial. An

Figure S.1 A 68000 based multibus compatible single-board computer (SBC) (*courtesy* Data Translation Ltd)

unmanned submersible craft developed by an international consortium to assist in the recovery, repair and replacement of damaged submarine cables buried in the seabed.

SCOPE OscilloSCOPE. An instrument that is able to provide visible images of single or multiple rapidly-varying electrical quantities. Images can be displayed with respect to time, or with respect to other electrical quantities. The most common type is the cathode-ray oscilloscope (CRO).

SCR (1) Silicon-Controlled Rectifier. A pnpn (pnpn) device constructed of a chip (CHIP) of four layers of semiconductor material that form three p-n junctions (p-n). In operation the forward anode-cathode current (I) is controlled by a signal being applied to a third gate electrode. A basic SCR is shown in *Figure S.2*. *See* CSR.

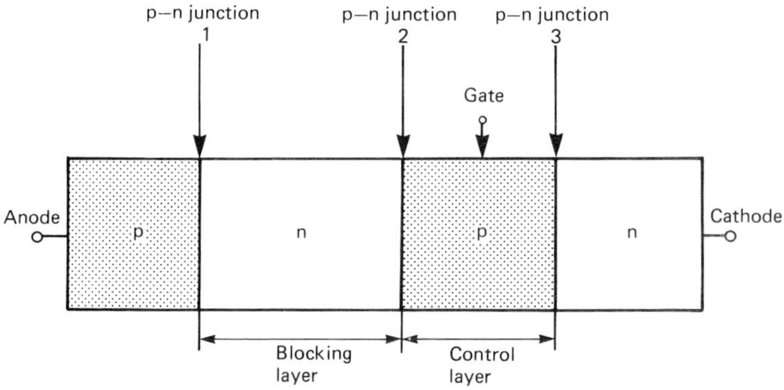

Figure S.2 A basic silicon-controlled rectifier (SCR)

(2) Sequence-Control Register. A specific storage location or register that holds the location of a specified item in a memory, containing the next instruction to be executed.

(3) Short-Circuit Ratio. The ratio of the field excitation for rated voltage (V,2) on open circuit, against the field excitation for rated armature current (I) on short circuit, of an alternator that is running at a rated frequency.

SCRAM Static Column dynamic Random-Access Memory. A dynamic random-access memory (DRAM) organized as 64K × 1 using one transistor cell arrays and housed in a 16-pin package. The memory on the same row address can be accessed only by selecting column address without the aid of a clock, as on a static random-access memory (SRAM). In this mode, cycle time can be equal to an access time while a DRAM usually requires a cycle time of twice the access time. Therefore, the access time becomes competitive.

SCS Silicon-Controlled Switch. A four-layer device similar to a thyristor but having both gate leads brought out. A basic SCS is shown in *Figure S.3*. It can be switched on by either a positive pulse to its cathode gate or a negative pulse to its anode.

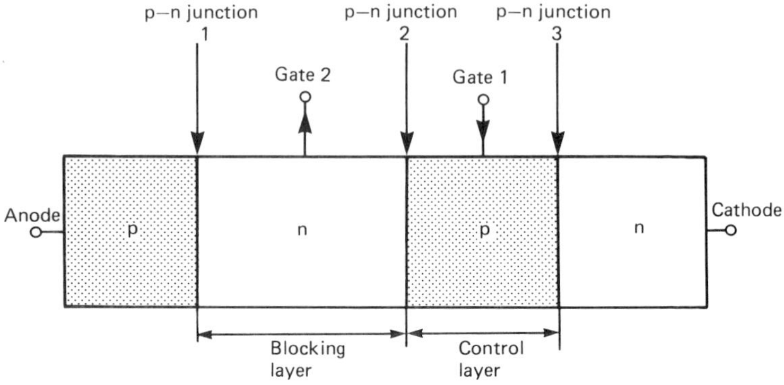

Figure S.3 A basic silicon-controlled switch (SCS)

SDA (1) Source Data Acquisition. Information being entered directly into a computer at the location where the data originates. An example of this is a point-of-sale (POS) system. (2) Source Data Automation. The coded recording of information on tags, paper tape (PT) or punch cards, which can be used repeatedly without rewriting being necessary in order to produce other records of data.

SDC Synchro to Digital Converter. A device that converts the three synchro voltages (*see* CX) into a binary digit (BIT) number, which can then be converted into an analogue voltage by a digital-to-analogue converter (DAC) or read directly by a computer.

SDF Supergroup Distribution Frame. Used in a frequency-division multiplexing (FDM) system, this frame provides for the interconnection of equipment and apparatus which transmit signals having frequencies in the 60-channel basic supergroup range.

SDLC Synchronous Data Link Control. A uniform discipline employing synchronous data techniques to transfer data between stations in a multipoint, point-to-point or loop configuration.

SDM (1) Space-Division Multiplexing. The combination of a number of separate wires or fibres into a bundle or cable, so that each fibre or bundle can be used as an independent communications path, channel (CHNL(d)) or set of channels. Typically a multiplexing arrangement might employ time division multiplexing (TDM) on each SDM pair in an optical cable. *See* OSDM, optical space-division multiplexing.

(2) Selective Dissemination on Microfiche. In response to subscriber's indicated areas of interest, this service provides them with microfiche copies of documents. *See* COM, computer output microfilm (COM).

SDR Statistical Data Recorder. A facility that logs the cumulative errors status of an input/output (I/O) device of a disk-operating system (DOS) on the system recorder file.

SDS/SDR Signal-Dependent Stereo/Signal-Dependent Response. A unique integrated circuit (IC) that monitors signal strength and in turn adjusts both (CHNL(g)) separation and high note response to match. It helps eliminate noisy stereo and sudden changes from stereo to mono.

Se Selenium. A semiconductor element of atomic number 34. Its grey allotrope form is very light-sensitive and is used extensively in photoconductive photocells.

SE (1) Sign Extend. Used while multiplying or dividing during a shift, this operation ensures that the binary digit (BIT) shifted is identical to the sign bit.

(2) Stop Element. The last element in each character used in start-stop transmission that has a minimum (MIN) duration assigned to it. During the transmission the receiving equipment is placed at rest to prepare for the reception of the next character.

SEAB SEriAl Bus. A coupling procedure based on the International Standards Organisation (ISO) reference model, which allows real-time data transmission at different defined levels and guarantees the autonomy of the automation structures, physical transmission system and the coupling procedure itself.

SECAM Système En Couleurs À Mémoire. A line-sequential colour television (TV(a)) system originating in France and used in several other countries including the USSR. In operation the signal which conveys information about the luminance of the television picture is transmitted by amplitude modulation (AM) of the vision carrier. Information concerning colour is transmitted by frequency modulation (FM) of the vision subcarrier. Colour saturation and hue (chrominance) signals are sent separately on alternate lines.

SELFOC SELf-FOCusing optical. A frequency selective optical fibre (OF) that can focus its output (O,2) transmission wave by controlled geometry, refractive indices and other factors.

SEM Scanning Electron Microscope. An electron microscope with a beam that can scan an object field in a controlled or regular manner. It has many industrial uses; for instance it can be employed to expose resist material in high-resolution scanning patterns in the fabrication of integrated optical circuit (IOC) components.

SEMCOR SEMantic CORrelation. A system of indexing that is computer aided.

SERC Science and Engineering Research Council. Formerly the Science Research Council (SRC), this UK organization funds a number of research programmes into various aspects of science, engineering and computing.

SERT Society of Electronic and Radio Technicians. A professional UK institute that provides services for technicians and technician engineers in all branches of electronics.

SF Skip Flag. Typically a 1-binary digit (BIT) memory location in a microcomputer that represents a true or false skip or no-operation (NOP) condition, with respect to the instruction being executed by the processor.

SF_6 Sulphur HEXaFluoride. SF_6 is a heavy chemically inert, non-toxic, non-flammable gas, which is odourless and colourless. It was first produced as long ago as 1904; however, it was some 30 years later that American investigators found its most important property, that of being an excellent insulator and having electric-arc quenching abilities.

SFCS Slats and Flaps Control System. An intelligent avionics system, used for controlling the powerful slats and flaps of an airliner with safety. Its design incorporates two different types of 8-bit microprocessor (MP) independently programmed to ensure that the equipment is fail/safe.

SGN SiGN. An arithmetic function in beginners' all-purpose symbolic instruction code (BASIC). In operation it substitutes −1 if the value of x is less than 0, or 0 if x is equal to 0 and 1 if x is greater than 0. For example:

SGN (9.7) = 1
SGN (0) = 0
SGN (−6.7) = −1

S/H Sample and Hold. An analogue circuit that is capable of capturing and retaining a signal, so that it may be converted by an analogue-to-digital converter (ADC).

SHA Special Handling Area. An area for handling static sensitive devices (SSD) that is essentially free of electrostatic charge. The environment in an SHA is controlled in such a manner as to cause any electrostatic discharge to decay. This is achieved by the use of conductive materials for furnishings, tools and fittings, which can be augmented by the passing of ionized air across the working area. The general layout of an SHA is shown in *Figure S.4*.

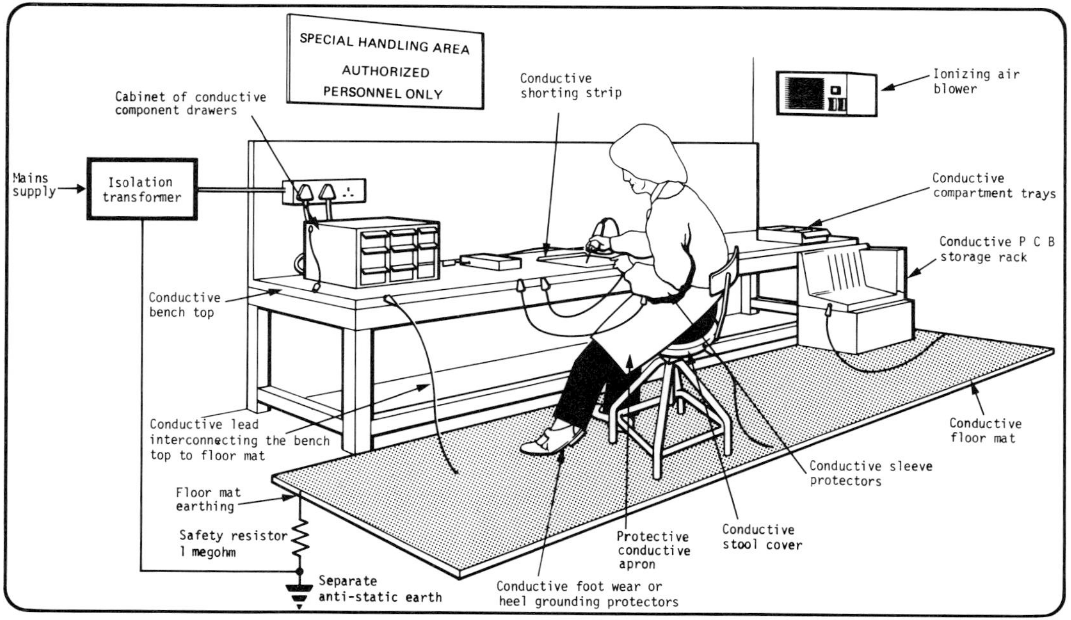

Figure S.4 The general layout of a special handling area (SHA)

SHF Super High Frequency. A frequency between 3 and 30 gigahertz (GHz) forming part of a larger continuous range of frequencies. SHFs are considered by international agreement to be radio frequencies in band number 10 and have wavelengths between 1 and 10 cm.

SHORT SHORT circuit. An abnormally relatively low-resistance connection between two points on a circuit (CCT). Such a condition may be accidental or unintentionally caused, for example when two wires touch that under normal conditions are insulated or separated from each other.

Si SIlicon. A non-metallic element. This abundant material is an intrinsic semiconductor and is one of the most important materials in the manufacture of transistors and other complex integrated devices (IC).

SI (1) The Système International d'Unités (International System of Units). A coherent system of units based on the metre, kilogram, second, coulomb and dependent units on which international agreement has been reached by the General Conference on Weights and Measures. The three classes of units are *base units, derived units* and *supplementary units.* There are seven base units: the metre (m), the kilogram (kg), the second (s), the ampere (A), the kelvin (K), the candela (cd), and the mole (MOL). Derived units can be formed by combining base units. Thus, the unit of force (the newton, N) can be produced by combining the first three base units. Two supplementary units at present defined are the radian and the steradian, which are the units for plane and solid angles respectively. Prefixes are shown in the table below.

SI Prefixes and their related Multiplication Factors and Symbols

Multiplication factor		*Prefix*	*Symbol*
1 000 000 000 000 000 000	$= 10^{18}$	exa	E
1 000 000 000 000 000	$= 10^{15}$	peta	P
1 000 000 000 000	$= 10^{12}$	tera	T
1 000 000 000	$= 10^{9}$	giga	G
1 000 000	$= 10^{6}$	mega	M
1 000	$= 10^{3}$	kilo	k
100	$= 10^{2}$	hecto	h
10	$= 10^{1}$	deca	da
0.1	$= 10^{-1}$	deci	d
0.01	$= 10^{-2}$	centi	c
0.001	$= 10^{-3}$	milli	m
0.000 001	$= 10^{-6}$	micro	μ
0.000 000 001	$= 10^{-9}$	nano	n
0.000 000 000 001	$= 10^{-12}$	pico	p
0.000 000 000 000 001	$= 10^{-15}$	femto	f
0.000 000 000 000 000 001	$= 10^{-18}$	atto	a

(2) SuperImpose. A process used in computer operations where data are moved from one location to another. Characters or binary digits (BITs) are superimposed onto the contents of a specified location.

(3) Serial Input. A method of data transfer between a computer and a peripheral device where information is transmitted for input (I/P) to the computer, binary digit (BIT) by binary digit over a single circuit.

(4) Sample Interval. The discrete period of time (regular or irregular) over which the process of obtaining the value of a variable is made.

(5) Shift-In character. A control character that is able to bring a standard character set back into operation. It reverses the effect of the shift-out character (SO).

SID (1) Sudden Ionospheric Disturbance. Unexpected radio fade-out caused by sudden short duration and abnormal ionization increases in the ionosphere D and E layers.

(2) Sound Interface Device. An integrated circuit (IC) found in some microcomputers. Used to generate sound effects.

SIGLE System for Information on Grey Literature in Europe. A system for detecting, identifying and collecting widely available although unpublished literature. It is sponsored by the Commission of European Communities.

SIL (1) Single In-Line. A concept that provides standardized location and connection of components and circuits, on the basis of a row of connections on a 0.1 inch pitch. The term can also be applied to the pins on a single in-line package (SIP) or to the holes in a socket or printed circuit board (PCB).

(2) Systems Implementation Language. A term that can be applied to a number of different high-level computer programming languages. These can either produce very efficient code, or have other particular features that enable them to be used for writing systems programs.

SIMD Single Instruction stream, Multiple Data stream. A particular type of design and construction of a computer. It can include in its make-up vector machines, parallel processors having one control unit and a number of arithmetic and logical units (ALUs), as well as some associative processors.

SIN SINe (a) A mathematical term defined as the line drawn on the end of the arc perpendicularly to the radius through the other end. Often used in the context of the sinusoidal wave, being a periodic oscillation of a given amplitude. *See* alternating current (AC,1). (b) A standard Pascal (*see* PASCAL) function used to plot a sine curve.

SINAD SIgnal to Noise and Distortion. The comparison of a signal against noise and distortion, expressed as a ratio and measured in decibels (dB).

SiO_2 Silica. An abundant natural compound that occurs in a number of forms including quartz and sand. It is an important source of the silicon that is used for the manufacture of electronic devices, components, integrated circuits (ICs) and as natural quartz for its piezo-electric properties.

SIO Serial Input/Output. A method of data transfer between a computer and a peripheral device. The data are transmitted for input (I/P) to the computer or out to a peripheral device, binary digit (BIT) by binary digit over a single circuit.

SIP Single In-line Package. A package to house networks, components or circuits, with one row of connections which extend down at 0.1 inch intervals.

SIPO Serial In Parallel Out. A complex structure of shift registers in a large-scale integrated (LSI) circuit which is created using either flip-flop (FF,2) elements for static operation or simple clocked (CLK) invertors without feedback.

SIR Selective Information Retrieval. The access of particular items of information held on computer files, in response to requests made by users indicating a particular area of interest.

SIS Semiconductor-Insulator-Semiconductor. Typically a solar cell constructed in a sandwich form, consisting of a transparent conductor such as the oxides of tin, indium, zinc or cadmium for the top contact layer. The second layer is a very thin insulating layer and the bottom layer is a p-doped (p-type) semiconductor. The top conducting oxides are actually heavily doped semiconductors; hence the resulting device is known as an SIS solar cell.

SISD Single Instruction stream, Single Data stream. A type of design and construction of a computer that is accepted as the normal or conventional make-up.

SITA Société Internationale de Tèlècommunications Aeronautiques. An organization that operates a message and data network for a large group of airline companies.

SL Single Lead. An SL cable is a three core (three conductor) cable that has each core sheathed with a single layer of lead as a covering.

SLAC Subscriber Line Audio processing Circuit. The base for analogue/digital (A/D) conversion to a subscriber's line in an electronic telephone exchange. *See* SLIC.

SLANG Systems LANGuage. A computer programming language in which statements normally correspond directly with machine level instructions and conversely, each machine operation is matched in a high level language statement.

SLD Super-Luminescent Diode. A light-emitting diode (LED) with a narrow spectral width and a high intensity of electromagnetic radiation. It is normally used as a source for optical fibre (OF) transmission systems. It exhibits both spontaneous light emission (as in a LED) and stimulated light emission (as in a LASER) simultaneously. Its output (O,2) has a narrow spectral width and a higher radiance than an ordinary LED.

SLED Surface-Emitting Light-Emitting Diode. This type of light-emitting diode (LED) transmitter, emits light from its top face and a microlens collimates this radiation, efficiently coupling to a fibre optic (FO).

SLIC Subscriber Line Interface Circuit. This adapts the subscriber's line to the exchange and provides the telephone with current (I), drives both ring and test relays and supervises the line. *See* SLAC.

SLICE Selective Line Insertion Communication Equipment.

SLIC/SLAC Subscriber Line Interface Circuit/Subscriber Line Audio processing Circuit. These contain all the line functions required by an electronic telephone exchange.

SLSI Super-Large-Scale Integration. Solid state circuitry usually on a single chip of silicon that has typically 100 000 or more transistors per chip.

SLT Solid Logic Technology. Technology for the production of microelectronic circuits known as logic circuits because they transmit and control electrical impulses representing information within computing devices. The tiny devices operate at speeds which are as fast as six thousand millionths of a second (s).

SM Surface Mount. A system that enables connections to be made on the surface of printed circuit boards (PCBs). It is ideal for stacking feed through connectors for PCB applications.

SMART System for the Mechanical Analysis and Retrieval of Text. Designed for the interactive reviewing or searching of full text documents, this system uses particular words and phrases to search. It analyses the text to produce a list of documents, placing those incorporating the search words at the top.

SMD Surface Mounted Devices. The surface mounting (SM) of components is used as an efficient production method for populating printed circuit boards (PCB). SMDs are designed with solder pads or short leads and are much smaller than comparable leaded components. In contrast to conventional components, the leads of which must be inserted into holes, SMDs are attached directly to the surface of the PCB and then soldered.

SMF Systems Management Facilities. Available on various

machines, this group of utilities makes possible the management of the particular system for tasks such as counting routines.

SMPS Switch Mode Power Supply. A power supply unit (PSU) that employs a switching technique. Its implementation provides considerable savings in size, weight and efficiency.

SMPTE Society of Motion Pictures and Television Engineers (USA).

SMT Surface Mount Technology. The scientific and technical activities, research and development (R&D), applied towards the development of surface mount (SM) products, systems or services.

SMX Semi Micro Xerography. Producing xerographic copies of documents, this technique enables micrograph forms of documents to be input into a computer by means of a special reader.

S/N Signal-to-Noise ratio. *See* SNR.

SNA Systems Network Architecture. A particular design and construction for a distributed system, allowing computation of jobs at a number of physically separated locations. It also permits a number of devices to be interconnected in a variety of ways.

SNAPSHOT A colloquial expression for a dynamic dump of the contents of a particular memory storage location and register, which are being employed during a program execution.

SNOBOL StriNg-Oriented symBOLic language. A computer language developed by Bell Telephone Laboratories which has significant applications in program compilation and in the generation of symbolic equations.

SNR Signal-to-Noise Ratio. Usually expressed in decibels (dB), this is a comparison of the magnitude of a signal to that of noise (unwanted disturbance) at the same point in a circuit. The form of its measurement and expression can vary depending upon the type of signal and noise concerned.

SO (1) Shift-Out character. A character that is used to extend a standard character set by substituting it with another character set. Normally this is done to provide access to additional graphic characters.

(2) Small Outline. A method of miniature packing which is well suited to surface mounting (SM). It offers space advantages over comparable dual-in-line packages (DIP), and its shorter leads are flexible, thus compensating for mechanical stress. Also known as Minidip. *See Figure S.5.*

SOC State Of Charge. A meter that indicates the percentage of usable charge remaining in a battery (BAT).

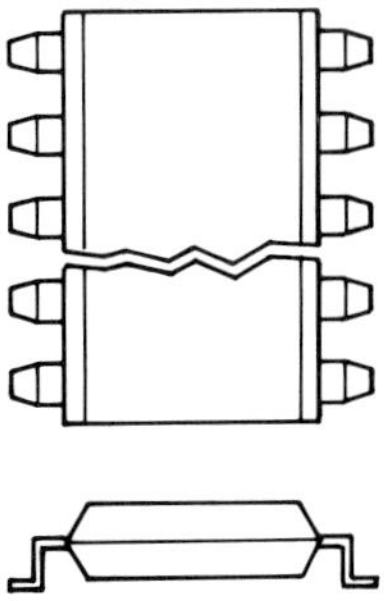

Figure S.5 Detail of a typical small outline (SO) packaged, surface mounted device (SMD)

SOD Small Outline Diode. This surface mounted device (SMD) consists of a diode (D,5) housed in a small outline (SO) cylindrical package, similar to a metal electrode face bonding (MELF) package.

SOF Shorts, Opens & Fixture. A preliminary check undertaken by automatic test equipment (ATE) to verify proper contact between the ATE pin electronics and the board, and to screen for shorts (SHORT) or open artwork.

SOH Start Of Header. A control character used in communications that is placed at the start of a sequence of characters comprising a machine sensible address or routing information. The sequence of characters is referred to as a 'header'.

SOIC Small Outline Integrated Circuit. A surface mounted device (SMD), consisting of an integrated circuit (IC) housed in a small outline (SO) package.

SOLINET SOuth-eastern LIbrary NETwork. A network connecting together over 200 libraries in the SE of the USA, thus permitting them to share facilities such as data processing (DP,1) and access to bibliographic information.

SOM (1) Start Of Message. A single character or group of characters used in telecommunications by a terminal to provide indication to other stations that the following transmitted information contains the addresses of those who should receive an answering message.
(2) Self-Organizing Machine. A machine that conducts its own internal organization, requiring no external intervention.

SOP Standard Operating Procedure. A particular mode of operation that is regularly or commonly used.

SOS Silicon-On-Sapphire. Materials in layers that produce devices capable of achieving bipolar speeds by using metal-oxide semiconductor (MOS) technology, achieved by insulating the circuit components from each other. SOS can also refer to the process of fabricating these devices.

SOT Small Outline Transistor. This surface mounted device (SMD) consists of a transistor housed in a small outline (SO) style of package. *See Figure S.6.*

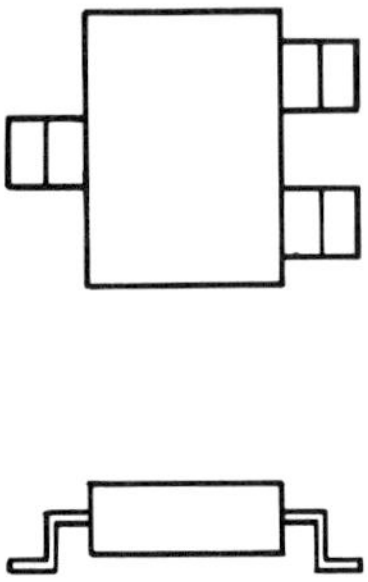

Figure S.6 Detail of a typical small outline (SOT) packaged, surface mounted device (SMD)

SP SPace character. (a) A character used in data processing and telecommunications. It is a non-printing character that is employed as a blank to separate words, etc. (b) An area in a memory that consists of blank characters. (c) A binary zero condition on a channel (CHNL(c)) or line during data communications.

SPADE Single-channel-per-carrier, Pulse-code-modulation, multiple-Access Demand-assignment Equipment. A system that has been developed for use in association with INTELSAT IV satellites to make more efficient use of their transmission capacity. It allows each carrier to be transmitted only when actually required for speech, enabling the transmitter/responder (TRANSPONDER) in the satellite to handle a much larger number of circuits before becoming overloaded.

SPC Stored Program Control. A program held in an electrically alterable memory or store that contains appropriate instructions to effect a method of control. SPC is often used to describe processor-controlled telephone switch systems that are controlled in the same manner, although differing widely in construction.

SPDT Single-Pole Double-Throw. A construction of switch that can be incorporated into a single line or pole. It has a single centre or common terminal with a movable blade. When activated (thrown) the blade is able to make contact, in one direction or the other, with one of two available terminals.

SPEC SPECification. An approved or accepted requirement or criterion for controlling technical performance or practice. It is by necessity precise, thoroughly developed and written down, updated to conform with current practices and used widely in order to be most effective.

SPG Synchronizing Pulse Generator. A unit used to produce master timings for studios etc.

SPICE Simulation Program with Integrated-Circuit Emphasis. A program used for the analysis of integrated circuits (ICs). It is a general purpose circuit simulator which can be used for three main analysis techniques: (i) linear alternating current (AC,1) analysis, (ii) non-linear direct current (DC,1) analysis, and (iii) non-linear transient analysis.

SPIN Searchable Physics Information Notes. A store of publications of the American Institute of Physics held on computer files.

SPL (1) System Programming Language. Developed for writing systems orientated software packages, this computer language's syntax and semantics are used to produce small and fast loading modules for specific equipment.

(2) SPLice. The action of joining two or more cables together providing suitable protection for the joint or joints such that the final assembly is as close as practical to a continuous cable. It is also used to describe a completed joint.

(3) Sound Pressure Level. A measure parameter, it is the direct sound pressure produced by a loudspeaker over its full frequency range, at a distance of 1 metre (m,2), for a specified power input.

SPRAT Small Portable RAdar Torch. A lightweight, about 2.5 kilogram (kg), hand-held radar system that is based on a Gunn diode and has an effective range of around 600 metres (m,2).

SPS (1) Symbolic Programming Systems. A computer programming language that represents terms by quantities and locations.

(2) String Process System. A package of computer software that has been designed and assembled for the manipulation and processing of consecutive sets or strings of characters.

SPST Single-Pole Single-Throw. A construction of switch that can be incorporated into a single line or pole. It has a single terminal with a moveable blade which is able, when activated (thrown), to make contact with another single terminal.

SPX SimPleX. Pertaining to a circuit or channel (CHNL) that permits operation or transmission in one direction only, as opposed to duplex (DX).

SQA System Queue Area. Maintained by an operating system (OS), this is a main memory storage area specifically for control, status information or tables.

SQR SQuare Root. A mathematical term, the SQR of a given number is the number of which it is the square, i.e. the SQR of 9 is 3 and of x^6 is x^3. It is represented by the sign $\sqrt{\ }$.

SQUID Superconducting Quantum Interference Device. A device produced by using conventional thin-film techniques, from superconductors such as Niobium, *see Figure S.7.* In its analogue mode it will detect magnetic fields down to as low as 10^{-17} tesla. In its digital applications, it combines unprecedented high switching speeds with low dissipation for 100 GHz binary dividers, as well as uses in very-large-scale integration (VLSI) gate and memory arrays for a superconducting computer which is currently being investigated by International Business Machines (IBM).

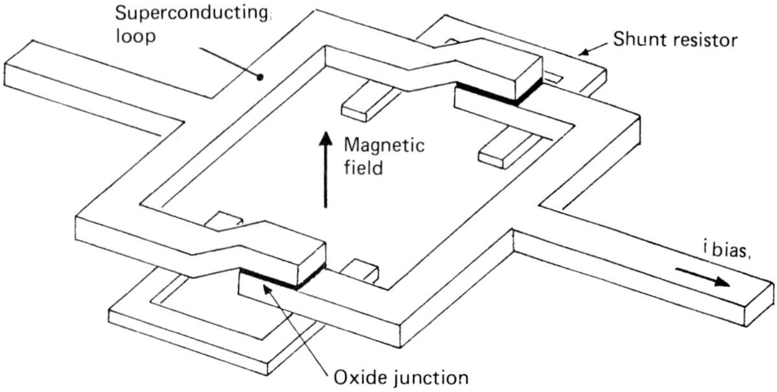

Figure S.7 An example of a planar thin-film superconduction quantum interference device (SQUID)

S-R Set-Reset memory. The simplest form of memory or flip-flop, which can be implemented in relays by a latching circuit. It consists of two cross-coupled NAND or NOR gates and stores a binary digit (BIT) controlled by the signals applied to the two inputs. Inputs are S and R, and the outputs are Q and $\bar{Q}$. *See Figure S.8.* (*See* RS.)

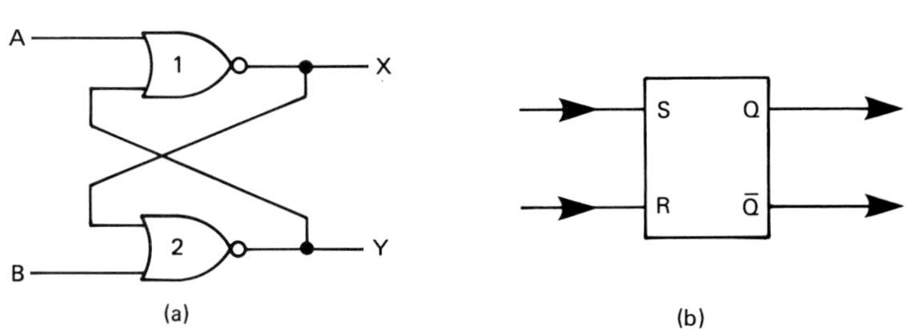

Figure S.8 (a) The S-R logic diagram showing two cross-coupled NOR gates. (b) The S-R flip-flop symbol

SRAM Static Random-Access Memory. A random-access read/write memory (RAM) that is able to retain data indefinitely so long as it has an adequate power supply. A SRAM can be implemented in either bipolar or metal-oxide semiconductor (MOS) and is basically a read/write memory that stores information or data in integrated flip-flops (FF).

SRBP Synthetic Resin Bonded Paper. A laminated sheet of insulating material consisting of paper bonded together with synthetic resin under the action of pressure and heat. It is available in a number of grades in thicknesses from 0.2 to 50 mm, and is widely used for circuit boards, panels and terminal boards.

SRC Science Research Council. Now known as the Science and Engineering Research Council (SERC), this organization funds a number of research programmes into various aspects of science, engineering and computing.

SRCNET Science and engineering Research Council NETwork. A packet-switched data transmission network in the UK run by the Science and Engineering Research Council (SERC).

SRIM Selected Research In Microfiche. An information service held on microfiche that can provide document reference concerning new material.

SRL Structural Return Loss. A measure of the reduction in power (loss) suffered by coaxial cables due to various irregularities in their structural geometry which in turn gives rise to forward and backward echoes.

SSB Single SideBand. A transmission method where either of the frequencies above or below the carrier of an amplitude modulated (AM) wave is used, the unwanted frequency (or sideband) as well as the carrier being filtered out.

SSC Station Selection Code. A code employed to indicate to a computer that access from a specific terminal or data processing (DP,1) station is required by the user.

SSD Static Sensitive Device. An electrical or electronic device that can be irreparably damaged as a result of electrostatic charges, usually via its terminals. SSDs can include metal-oxide semiconductor (MOS) devices, junction field effect transistors (JFETs), beam lead diodes, hybrid integrated circuits (ICs), panel electronic circuits (PECs), some bipolar transistors and other small junction area devices.

SSI Small-Scale Integration. The design and production of integrated circuits (ICs) or microcircuit chips (CHIPs). The solid state circuitry is usually produced on a single chip of silicon having fairly simple circuits with fewer components in a given area than medium-scale integration (MSI). Typically an

SSI circuit contains from one to four logic circuits.

SSR (1) Solid State Relay. A relay having all its constituent parts made from solid state components and involving no mechanical movements. The lack of physical contacts and no moving elements increases the mean time between failure (MTBF), decreases problems such as radio frequency interference (RFI), provides resistance to corrosion, enables usage in explosive environments and provides rapid switching speeds.
(2) Secondary Surveillance Radar. An essential facility for aircraft identification and control normally employed as an integral part of air traffic control systems. It is a co-operative system in which civil aircraft are fitted with a transmitter/ responder (TRANSPONDER), which is able to reply with a coded signal to an interrogation signal made by the SSR.

SST Single-Sideband Transmission. An amplitude modulation (AM) transmission system in which only one of the sidebands (frequencies above and below the amplitude modulation carrier wave) is transmitted, the unwanted carrier and sideband being suppressed.

STAR Serial Transmitter And Rom. A combined message store, i.e. read-only memory (ROM) and transmission unit. Designed to be used in conjunction with a series of alphanumeric readouts, it can inform machine supervisors in plain English what the plant under their control is doing. The primary application is for fault identification. *See Figure S.9.*

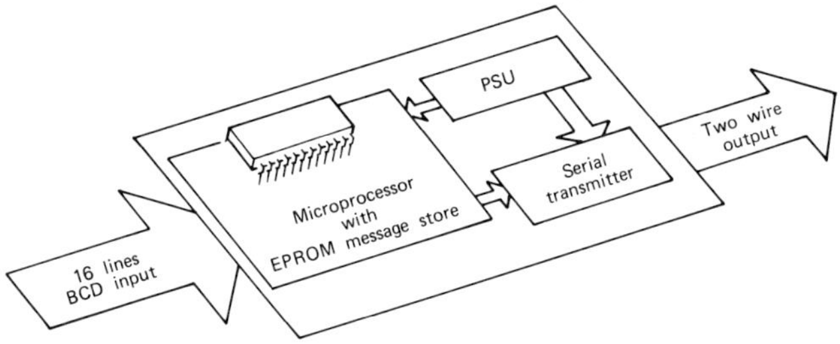

Figure S.9 Block diagram of a serial transmitter and random-access memory (ROM)

stat- A prefix of a unit used in the now obsolete centimetre, gram and second (CGS) electrostatic system of units.
1 stat ampere = 3.336×10^{-10} ampere (A)
1 stat volt = 2.998×10^{2} volts (V)
1 stat ohm = 8.988×10^{11} ohms

STD (1) Subscriber Trunk Dialling. A form of service provided by a telephone exchange in which subscribers are able to call other subscribers outside their own local area without operator

assistance. All the UK is now on STD. It is also known as 'direct distance dialling' (DDD).
(2) STanDard. An approved or accepted requirement or criterion for controlling technical performance or practice. It is by necessity precise, thoroughly developed and written down, updated to conform with current practices and used widely in order to be most effective.

STDM Synchronous Time Division Multiplexor. Able to share a synchronous data link, an STDM is a multiplexor (MUX) that scans and interleaves binary digits (BITs) into the frames of the incoming data.

STI Scientific and Technical Information. A generic term used to describe data of a scientific or technical nature.

STL Schottky Transistor Logic. A form of integrated injection logic (I^2L) where the collectors are formed as Schottky barriers to reduce any excessive storage of charge at the collector junction. This results in an increased speed of operation as well as a reduction in the total voltage (V,2) swing.

STP Standard Temperature and Pressure. Used in physical measurement these reference quantities or values are zero degrees celcius or centigrade (C,8) for temperature and 760 millimetres (mm) of mercury or 101 325 pascal (Pa) for pressure. Also known as 'normal temperature and pressure'.

STRESS STRuctural Engineering System Solver. A computer language employed in civil engineering for finding solutions to structural analysis problems.

STRUDL STRUctural Design Language. A computer language, developed from the structural engineering system solver (STRESS) language, which is employed in the analysis and design of structures.

STTL Schottky Transistor-Transistor Logic. A transistor-transistor logic (TTL) device that employs Schottky diodes for an improved speed of performance.

STX Start of TeXt. A control character used in communications that is placed at the head of a sequence of characters (text) and is transmitted through to the ultimate destination. A heading sequence started by a start of header (SOH) can be finished by an STX character.

SUB SUBstitute. (a) A specific control character that is used to replace an invalid character. (b) The action of replacing a particular element of information with another element of information.

SUHL Sylvania Universal High-speed Logic. Introduced in the early 1960s by Sylvania in opposition to Texas Instruments' (TI) transistor-transistor logic (TTL) range. At the time there was considerable rivalry between the merits of the two series until the market decided in favour of TI.

SVC SuperVisor Call. An instruction that interrupts (INT) the program which is executing it, and switches control to an executive routine.

SVR Super Video Recorder. Grundig's particular format used for video cassettes and recorders.

SWA Single-Wire Armour(ed). When cables are not installed in conduit or trunking, they may require armour, most commonly provided by galvanized steel wire (GSW) helically applied for protection in a single layer, being known as single-wire armour.

SWAP Stewart-Warner Array Program. A chip (CHIP) of transistors and resistors manufactured with connections implemented in a user specified manner.

SWG British Standard Wire Gauge. A legal UK system that employs numbers to designate the diameters of wires.

SWOP AMP SWitchable input OPerational AMPlifier. A precision operational amplifier with two identical input stages sharing a common load and output stage, as shown in *Figure S.10.* One input at a time is selected using transistor-transistor logic (TTL), complementary metal-oxide semiconductor (CMOS) logic or user programmable logic levels. The unselected input circuitry exhibits a very high input impedance drawing less than one picoamp. Applications include extremely-low drift amplifiers, multiplexed input amplifiers with good matching, switchable gain circuits and auto zero systems.

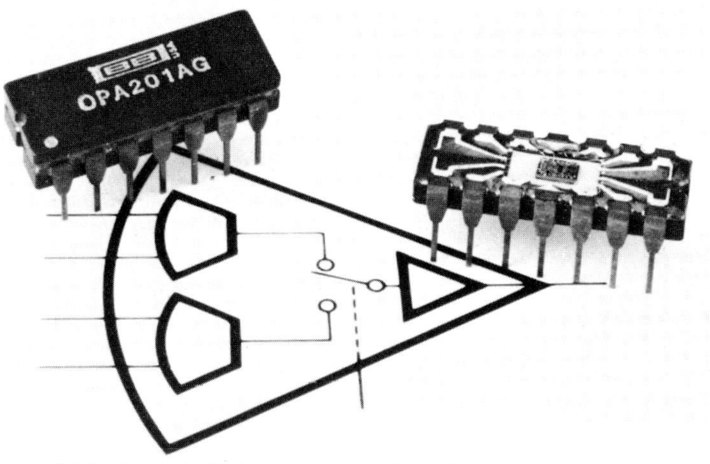

Figure S.10 A switchable input operational amplifier (SWOP AMP); this particular equipment has two inputs and a single output (*courtesy* Burr-Brown International Ltd)

SWR Standard Wave Ratio. The ratio of the magnitude of an anti-node compared to a node in a field pattern of a transmission line produced by two similar frequency waves propagating in different directions. Also known as Standing Wave Ratio.

SYN (1) SYNchronization signal unit. Used in common channel signalling (CCS), this group of information and check binary digits (BITs) contains distinctive information that can swiftly set up synchronism between sending and receiving terminals in a signalling network.
(2) SYNchronous idle character. A control character, used in synchronous data transmission, that is transmitted to set up synchronism between one data terminal and another. It can also maintain synchronism when no other character is being transmitted.
(3) SYNchronizing pulses. Pulses that are employed in the transmission of television (TV(a)) pictures to provide a stable reference frame.

SYSGEN SYStem GENeration. (a) Used in some computer systems to create a supervisor and delay management support for a particular system configuration. Also used to include the program products that have been ordered. (b) The action of producing an operating system (OS) by selection from a large number of modules. The units are modified to create the optimum system.

SYSLOG SYStem LOG. A data set that is capable of storing job-related information, operational data, descriptions of unused occurrences, commands and messages to or from an operator.

SYSTRAN SYStem TRANsatlantic. An automatic system of machine translation (MT,3) that provides translations between English, French, Russian and Spanish.

SXS Step-By-Step switch. A switch that moves in such a way that it is synchronized with a pulse device such as a rotary telephone dial. Every digit that is dialled causes the movement of successive selector switches which carry the connection forward until the desired telephone line is reached. It is also known as a 'stepper switch'.

T

2D TWO Dimensional. Refers to images etc. that have or appear to have length and breadth but no depth.

2 DEG TWO-Dimensional Electron Gas. The heterojunction in a transistor is structured such that high mobility carriers can be introduced into an extremely narrow channel (CHNL(h)). This high mobility results from the ionized donors being isolated from free carriers. This isolation is able to be increased with the introduction of an undoped layer which limits the vertical freedom of unrestricted movement in the planar direction. Such a fluid is referred to as a two-dimensional (2D) electron gas.

3D THREE Dimensional. Refers to images, object etc. that have or appear to have length, breadth and depth.

T (1) T antenna. Comprising a horizontal conducting element insulated at its ends with a down lead connected at its midpoint, this type of antenna may have an element made up from a number of conductors connected in parallel.

(2) T network. An unbalanced network circuit consisting of three impedances, two series elements and a shunt element connected between their junction and earth, as shown in *Figure T.1.* It is also known as a star or Y- network.

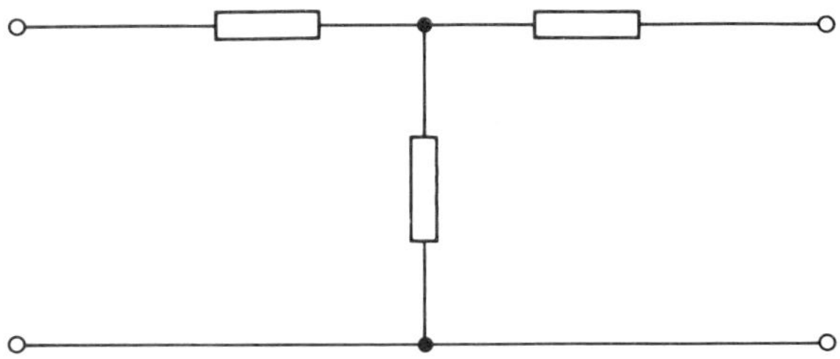

Figure T.1 A T-network

(3) A flip-flop (FF) circuit that has one input (sometimes two); when a pulse appears at the input(s) this will cause the circuit to change its state to one of two possibilities.

(4) Tesla. Unit of magnetic flux density, equal to one weber (Wb) per square metre (m,2) of circuit area.

Ta TAntalum. A metal of atomic number 73, having extremely high resistance to corrosion.

TAB (1) TABulation character. A control character that is employed to move a printing mechanism or cursor in either a vertical or horizontal direction.

(2) Tape-Automated Bonding. A system used to mount semiconductor chips (CHIP) onto flexible continuous printed circuits (PC,1), which speeds production and improves reliability. All the inner leads are brought into contact simultaneously with the bonding pads on the chip. *See Figure T.2.* The substrate comprises a continuous strip of Kapton plastics film, with sprocket holes for automatic advance and precise alignment. It is also known as 'gang bonding'.

TAC Telex Data Acquisition and Control.

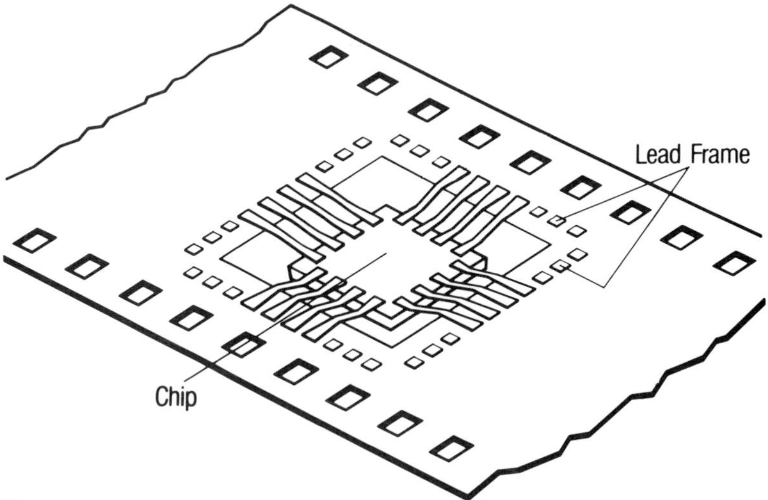

Figure T.2 The tape-automated bonding (TAB,2) process bonds leads simultaneously (*courtesy* Du Pont de Nemours International SA)

TACS Total Access Communications System. A cellular radio system in the UK which commenced service soon after midnight on the 1st January 1985. There are two system operators, Cellnet and Vodafone.

TASI Time Assignment Speech Interpolation. In a telecommunication voice circuit audible speech is present for approximately 45% of the time. TASI switching equipment can connect a party to an idle circuit while speech is taking place and disconnects the party when speech stops, so that a different party can use the same circuit. The efficiency can be improved from 45% to approximately 80%.

TAT TransATlantic cable. A submerged telecommunications cable. The designation of such cables are identified by digits (DGT,1), for example TAT-5.

TBC TimeBase Corrector. A short-term signal store that has been designed to reduce timing errors in video tape recorders.

TBMT Transmitter Buffer eMpTy. One of the five status binary digits (BITs) of a standard universal asynchronous receiver/transmitter (UART).

Tbps TeraBits Per Second. A signalling rate of 10^{12} binary digits (BITs) per second, that is one million bits per second.

TC (1) ThermoCouple. The symbol TC is used to represent thermocouple on parts lists and to annotate schematic

tre. A class 4 office where telephone operators are

present; if they are not, then it is called a toll point (TP). It is one of five class numbers assigned to offices according to their function in the US direct distance dialling (DDD) telephone network. Any one centre can handle telephone call traffic from one, two or more centres lower in the hierarchy, class 1 being the top classification.

TCAM TeleCommunications Access Method. A subsystem that controls the transfer of messages between an application program and remote terminals, providing a high-level message control language. It can be tailor made to customer application requirements.

TCB Task Control Block. A memory location containing information or a set of routines used to control or manage jobs or tasks within a computer system.

TCBM Time-Consistent Busy Hour. Beginning at the same time on each week day, a period during which the highest average telecommunication traffic is measured. A TCBM is an uninterrupted period of 60 minutes.

TCM Terminal to Computer Multiplexor. A unit that is able to control the use of several sending and receiving channels (CHNL(b)) between a terminal and a computer simultaneously.

TCO Thermal Cut-Off. Devices such as fuses which have been designed to provide fast, accurate protection against thermal overloads in electronic equipment, are considered to be TCOs.

TCP (1) Transmission Control Program. A program within a computer network that sets into motion formalized methods of communication between networks.
(2) Tape Conversion Program. A special computer program used for copying and/or converting paper tapes (PT) from one tape format to another, e.g. from hexadecimal (HEX) to binary format or from binary to hexadecimal format.

TCS Terminal Control System. Designed to manage multiterminal operations in a computer system, the TCS schedules the usage of hardware and all input/output (I/O) processing.

TCU Transmission Control Unit. An input/output (I/O) control unit that provides address information for messages to and from a number of remote terminals; it also acts as an interface between a central processor unit (CPU) and the terminals.

TCWB Tinned Copper Wire Braid. A concentric screen normally found just below the outer sheath of some cables. It is produced from a group of copper wire filaments, coated in tin to aid soldering and inhibit corrosion, that are interwoven cyclinderically to form a covering.

TD Transmitter-Distributor. A specific device in a teletypewriter

(TTY) terminal that makes and breaks the line in timed sequence. The term is generally used when making reference to paper tape (PT) transmitters.

TDM (1) Time Division Multiplex. A process, circuit (CCT) or device that enables the transmission of two or more signals over a common path by using successive time intervals for different signals.

(2) Time Division Multiplexing. A method of using a single path to provide a number of channels (CHNL(a)) by time-dividing the path into a number of time slots and then assigning each channel its own intermittently repeated time slot. At the receiving end, each time-separated channel is reassembled. Ideally suited for the transmission of digital (D,4) data the system is now used for digitized speech and other signals. Allocations of time-slots may be regularly repeated (*fixed cycle*), or may be made on demand (*dynamic*).

TDMA Time Division Multiple Access. A technique that allocates different time slots to various users and enables communicating devices at separate geographical locations to share a multi-point or broadcast channel (CHNL(a)).

TDR (1) Time Domain Reflectory. A system for making diagnostic measurements of fibre optics (FO) of new designs. It can detect and analyse variations of dispersion or dimensions along the fibre length. It is also known as a 'photoprobe' or 'radar'.

(2) Transmit Data Register. Employed in asynchronous communications interface adaptor (ACIA) transmission systems, this memory location is used to hold data that are ready to be transmitted out.

TEAM Terminology, Evaluation and Acquisition Method. A machine-aided translation (MAT) system located in Germany that is able to offer facilities in eight major European languages.

TeD TElDec. A video-disk system from Telefunken and Decca.

TED Transferred Electron Device. The generic name given to a family of bulk-effect (bulk negative resistance) devices centred around the Gunn-effect diode (D,5) and its mode of operation. They are fabricated from compound semiconductors such as gallium arsenide (GaAs).

TELE- A prefix that denotes distance; for example, television is vision transmitted over a certain distance; a teleprinter is a remote printer; telephone is sound transmission.

TELEBANKING TELEphone/TELEvision BANKING. Refers to the transfer of money between accounts via the telephone and or television, using a modulator/demodulator (MODEM) which communicates with a computer having details of an individual's account. *See* automatic teller machine (ATM) and personal identification number (PIN).

TELECOMM TELECOMMunication. Any process that allows information of any nature transmitted in a useable form (e.g. audible signals, visible signals, printed copy, fixed or moving pictures) to travel from a sender to one or more receivers. The transmission is achieved by means of any electromagnetic system (e.g. electrical transmission by wire, radio, optical means and guided waves).

TELEPUTER A combined TELEvision and COMPUTER. Designed to be the central component in the home telecommunications services, this unit combines six major technologies; colour television, videotex, computing, video cassette recording, video disk and telecommunications. The system can be set to access automatically any number of computers, because it has a built-in personal computer capability. Data gathered can be manipulated, analysed and processed at the terminal under the control of various access programs.

TELEX TELetypewriter EXchange service. An automatic dial-up dedicated switched network that provides for communication between teletypewriter (TTY) and teleprinter equipment worldwide. A terminal is used to produce information on a paper tape (PT) which is fed into a reader and then transmitted.

TEMP (1) TEMPerature. A measure of the average kinetic energy of the moving molecules or atoms in a material. Generally it is a measure of how hot or cold material is compared to a defined scale. The Système International de Unités (SI) unit of temperature is the kelvin (K), but the Celsius (C,8) or centigrade scale is normally used for everyday measurements.

(2) TEMPorary register. Typically a 12-binary digit (BIT) memory location that holds or latches an arithmetic and logical unit (ALU) operation, in order to avoid race conditions before it is sent to a final memory location.

TERA- A prefix denoting one million million, a billion (10^{12}); for example a terabit is 10^{12} binary digits (BITs).

TFT (1) Thin-Film Transistor. Fabricated using thin-film techniques, typical of a thin-film transistor, is an insulated-gate field-effect transistor (IGFET). It is manufactured on a substrate.

(2) Thin-Film Technology. A technology in which electronic circuits or elements are formed by a process in which vacuum-deposited or sputtered films are produced upon a supporting material. TFTs can be considered as films that are

less than 5 micrometres (μm) thick as opposed to films greater than 10 micrometres thick.

THD Total Harmonic Distortion. A measure of the total distortion produced in an audio system as a result of the production, within that system, of harmonics or multiples of the original audio signal.

THERMISTOR THERMally sensitive resISTOR. Usually fabricated from semiconductor material shaped as rod, bead or disk and named accordingly, thermistors are available in two forms: negative temperature coefficient (NTC) thermistors that have a resistance that decreases with increasing temperature, and positive temperature coefficient (PTC) thermistors that have a resistance that increases with increasing temperature. Both forms cover a range of temperature coefficient from around −65% +70% per degree celcius or centigrade (C,8).

THP Through (the) Hole Plating. Printed circuit boards (PCBs) having circuits on both sides which are connected electrically by the plating that constitutes the circuit tracks passing through holes in the board.

Ti TItanium. Resembling iron, this metallic element has an atomic number 22, atomic weight 47.90, specific gravity 4.5 and melting point of 1850 degrees celcius or centigrade (C,8).

TIMS Transmission Impairment Measuring Set. Normally a portable unit designed for in-the-field characterization of analogue circuits for data communications. The sets are usually compatible with International Telegraph and Telephone Consultative Committee (CCITT) standards and measure impulse noise, random circuit noise, level versus frequency distortion and signal-to-noise ratio (SNR).

TIP Terminal Interface Processor. A small computer connecting terminals to a packet-switching network.

TIPTOP Tape InPut/Tape OutPut. A method for introducing data to an input device or reading data from an output (O,2) device. The technique can include the use of plastics or metallic magnetic tape (MAGTAPE), perforated paper tape (PT) or fabric tape loops.

TKO TrunK Offering. An exchange facility permitting an operator to intrude into an established connection to offer one of the parties an incoming long-distance call.

T^{2}L Transistor-Transistor Logic. *See* TTL.

TLB Translation Look-aside Buffer. Currently used in 32-binary digit (BIT) processors having an on chip (CHIP) memory management unit, a TLB that helps simplify the process of translating a logical address to a physical address.

TLU Table Look Up. The action of finding a function value that corresponds to a stated or implied argument, the information being held in a table of function values stored in a computer.

TLX TeLeX. *See* TELEX.

TM Tape Mark. A special character written on a tape to signify the physical end of recording. It can indicate the actual end of the tape or it can be used to subdivide the tape into sections.

TMP TeMPerature. *See* TEMP.

TMS Transmission Measuring Set. Equipment made up of a calibrated signal generator and a level measuring set, that is used to measure the insertion loss or gain of a transmission path.

TNC Threaded-Nut Coupling. A type of connector used for coaxial cable (COAX) radio frequency (RF) interconnection. It consists of a coupling that employs a screw-on interface.

TNEP Total Noise Equivalent Power. A measure of the minimum detectable signal that can be found among noise in detection and multiplication processes. It can be defined as the amount of light falling on a photodetector, required to produce an output (O,2) signal equal to the noise output. It is usually expressed in nanowatts.

TNS Transaction Network Service. A data exchange system used in US metropolitan areas that provides a basic communications service for short data messages, normally enquiry-response financial transactions.

TO A designation for a package normally used to house discrete transistors. The enclosure is most often constructed from metal although some plastics versions are available.

TOD Time Of Day. (a) An electronic clock (CLK) device that is able to measure and record time in hours, minutes and seconds over a 24-hour cycle. When called by a central processor unit (CPU) it is able to provide the time. (b) A notation made to indicate the particular time of day an event occurred.

TOP Technical Office Protocol. A sister initiative to manufacturing automation protocol (MAP,4), begun by Boeing aimed at standardizing communications between computers operating in technical and office environments.

TOPSY Territorial OPerational SYstem. Sponsored by the Home Office, TOPSY is a management system designed for the Metropolitan Police, a stand-alone system with four data bases; personal records; manpower; overtime and crime for each divisional station.

TOS (1) Tape-Operating System. An operating system (OS) held on a tape or a number of tapes. Some systems have compilers and linkage editors on a system tape and use tape libraries for user programs and data files.

(2) Temporarily Out of Service. A designation given to a telephone subscriber's line when it is temporarily deliberately disconnected from equipment at the exchange.

TP Toll Point. A class 4 office where operators are not present; if they are then it is called a toll centre (TC). It is one of five class numbers assigned to offices according to their function in the US direct distance dialling (DDD) telephone network. Any one centre can handle telephone call traffic from one, two or more centres lower in the hierarchy, Class 1 being the top classification.

TPA Transient Program Area. An area in the memory map (MAP,2) of a microcomputer using a control program for microcomputer (CP/M) disk operating system (DOS). It allows the user to write programs to run so they can access areas of basic disk operating system (BDOS), basic input output system (BIOS) and console command processor (CCP) to use routines already present in the system.

TPI Tracks Per Inch. An expression of track density on a magnetic disk measured along its radius.

TPST Triple-Pole Single-Throw. A switch that can be incorporated into a triple line or pole. It has three terminals each having a moveable blade. The blades are so constructed that when activated (thrown), each simultaneously makes contact with its respective line terminal. *See Figure T.3.*

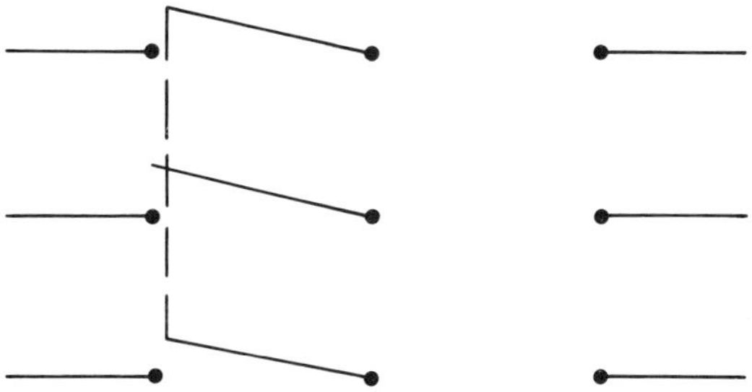

Figure T.3 Symbol for a triple-pole single-throw (TPST) switch

TR (1) TRansistor. The symbol TR is used to represent transistors on parts lists and to annotate schematic diagrams.
(2) Transmit-Receive. A switch employed in a radar (RADAR) system, where the system has a common antenna for both transmission and reception. It decouples the receiver automatically from the antenna during a period of transmission. *See* ATR.
(3) Transformer. *See* XMFR.

TRAN TRANsmit. The action of moving data or information from one location to another, normally via a transmission line or communications link.

TRANSCEIVER TRANSmitter/reCEIVER. A radio transmitter and receiver combined in a common enclosure employing some common circuit components for both transmitting and receiving, normally used for mobile or portable devices.

TRANSPONDER TRANSmitter/resPONDER. A device or circuit often used in aircraft and orbital satellites that is able to receive an input signal, act upon it and retransmit it. The device is used as a means of reaching a wide area with a radio transmitter.

TRAPATT TRApped Plasma Avalanche Triggered Transit. The TRAPATT diode (D,5) is a high efficiency microwave (μW) generator capable of operating in the gigahertz (GHz) region but in practice is limited to below the millimetre (mm) wavelength region.

TRF Tuned Radio Frequency. An amplitude modulated (AM) receiver that has one or more parts of its radio frequency (RF) amplification before its detector, and audio-frequency stages after. It has been superseded by superheterodyne receivers.

TRIAC TRIode AC semiconductor switch. A triac is an alternating current (AC,1) power circuit control device which can be used to control devices which can be used to conduct or block AC in either direction. Also known as the 'bi-directional triode thyristor'.

TRIAL Technique for Retrieving Information from Abstracts of Literature. Using the abbreviated representation of the contents of documents held on a computer file, this system is able to manipulate and process the information.

TRIAX (1) A colloquial expression referring to triaxial cable.
(2) TRIAXial cable. A cable similar to coaxial cable (COAX) which has an extra screen (making a total of three) used for television camera output. Often used for a camera with multiplexed outputs.

TRO Tough Rubber Sheath. A particular type of outer covering employed in flexible mains cable that have been designed for use in temporary building sites etc.

TROM Teletext Read-Only Memory. A dedicated read-only memory (ROM).

TS (1) Time Sharing. The use of a computer by two or more users simultaneously in real-time (RT,1). Ultra-high-speed switching of processing time is employed to achieve this sharing.
(2) Time Shared. The action of computer systems undertaking time sharing (TS,1).

TSC Transmitter Start Code. A code normally consisting of two letters that is used to call up a remote or outlying unit so that it automatically switches on its tape transmitter.

TSO Time-Sharing Option. An option that is made available with some operating systems (OS), to provide interactive use by a number of operators.

TSPS Traffic Service Position System. A stored program computer having telephone operator consoles that permit the efficient handling of calls requiring operator intervention.

TSS (1) Time-Sharing System. An interactive computer operating system (OS) that has been designed so that the central processor unit (CPU) can be simultaneously time shared (TS,2) amongst the users.
(2) Time-Shared System. A computer system that is capable of being time shared (TS,2) but has not been specifically designed for this purpose. *See* TSS,1.

TSWB Tinned Steel Wire Braid. A concentric screen normally found just below the outer sheath of some cables. It is produced from a group of steel wire filaments, coated in tin to aid soldering and inhibit corrosion, that are interwoven cylinderically to form a covering.

TT TeleTypewriter. *See* TTY.

TTL Transistor-Transistor Logic. One of the most commonly used logic families due to its versatility and high speed. It is a bipolar integrated circuit (IC) based on saturating transistors. The logic uses transistors with multiple emitters which are easy to fabricate, requiring only a single isolated collector region upon which a single base region and then several emitters are diffused. *See Figure T.4.* TTLs have high power consumption. Also known as T^2L.

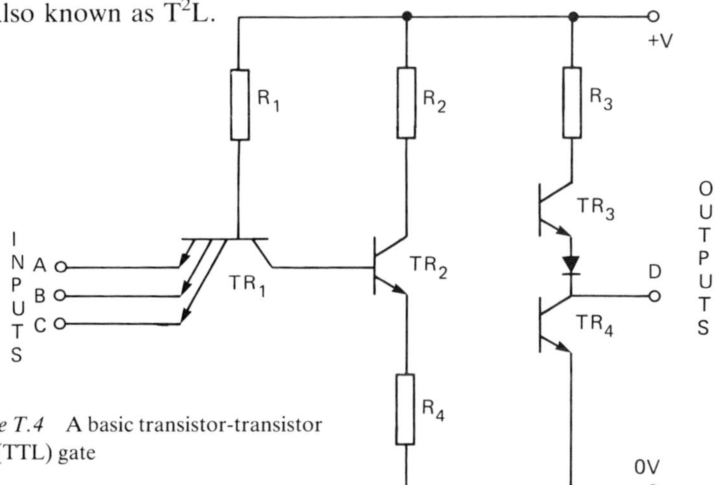

Figure T.4 A basic transistor-transistor logic (TTL) gate

TTS TeleTypeSetting. A system of typesetting remotely where a typesetting machine is controlled by input (I/P) from a punched paper tape (PT). This can be achieved over a communications link or directly.

TTY TeleTYpewriter. An electromechanical unit comprising a keyboard (KB,1) and a printer. Originally used to transmit information in communication systems, it is now used to communicate with computers using American Standard Code for Information Interchange (ASCII) or other codes.

T-type A specific, but little-used form of flip-flop, that has been designed such that its outputs (O/P) toggle on each low to high change of a clock (CLK) signal applied to its input. The JK flip-flop (*see* JK) can be implemented as a T-type.

TV TeleVision. (a) The technique of transmitting electrical signals in order to provide moving or stationary images with or without sound over a distance, via a suitable channel (CHNL(g)) or path. (b) A system that fulfils the techniques described in TV(a). (c) A receiver that is capable of receiving the signals produced by a system described in TV(b).

TVI TeleVision Interference. Interference to a domestic television (TV(c)) set can be caused by a local radio transmitter, or be produced by unsuppressed electromechanical devices such as electric motors.

TVT TeleVision Typewriter. A unit that is capable of displaying information on a television screen.

TWEETER A high frequency loudspeaker designed to operate only on high audio frequencies in high-fidelity (HI-FI) sound reproduction systems. It is usually part of a larger loudspeaker assembly that consists of two or more speakers. (*See* WOOFER.)

TWT Travelling Wave Tube. A hardware device that is employed to amplify microwave (μW) and other high frequency (HF) signals. Within the tube an electron beam interacts continuously with a radio frequency (RF) electromagnetic field to produce amplification or, in some types of tube, oscillation at microwave (μw) frequency.

TWX TeletypeWriter eXchange service. A North American public switched teletypewriter (TTY) service that has suitably arranged teletypewriter stations where customers are able to rent a teletypewriter provided with lines to a central office for access to other such stations throughout the USA and Canada. The service uses both American Standard Code for Information Interchange (ASCII) and Baudot machines and equipment.

TX (1) Transmitter. A unit or circuit in a communication system that converts information (often audible) into electrical signals

ready to be conveyed from one designated location to another by means of a transmission line, waveguide, radio channel (CHNL(a)) or wire. Normally on communications equipment a red panel indicator shows that a TX circuit is in operation.
(2) TeleX. *See* TELEX.

U

U (1) Rack height. A 'U' value is equivalent to 1¾ inches (44.45 mm) when referring to the height increments of a 19-inch racking system; for example a 3U-high rack would be equal to 5¼ inches. The standard was first set by the American Post and is now a *de facto* standard IEC2g. The U values can also be equated to the PCBs cards or other items that fit into the racks.
(2) Blue colour difference signal. The luminance signal employed in colour television (TV,1) transmission is the sum of all three colours. U is the colour difference signal (B-Y) modulated on the carrier, where B is blue and Y is the luminance signal. Two such signals U and V are employed to derive the third.

UA User Area. Part of a computer disk where semipermanent data may be stored. It can also be used to store programs, subprograms and subroutines.

UADS User Attribute Data Set. A collection of information held on a computer file concerning the users of a particular computer system.

UART Universal Asynchronous Receiver Transmitter. A particular device that is used to interface a data terminal or word parallel controller to a communication network.

UAX Unit Automatic Exchange. A UK designation given to an unattended automatic telephone exchange, that provides a service to a small community.

UBC Universal Block Channel. A direct memory access (DMA) channel (CHNL(b)) that operates in a block-mode, for high-performance peripheral controllers such as those used with disks or magnetic tapes (MAGTAPE).

UBL UnBLocking signal. A signal that is transmitted to a busy-out circuit (CCT) at its distant end. The UBL ends the busy signal, which may be continuous, indicating that the circuit is again available for use.

UC Upper Case. Capital letters of any type face. (*See* LC,1.)

UCLA UnCommitted Logic Array. Consisting of partially processed wafers of standard chip (CHIP) design, the UCLA provides a compromise between the standard and custom device, as the customer is able to specify the interconnections that are necessary to provide the particular functions required.

UCS (1) Universal Character Set. A printing device with a UCS facility can allow a number of character sets to be used, thereby providing any standard typeface.
(2) User Control Store. A section of a memory circuit designed to hold the sequence of commands which determine operations. It is set aside for the user's microprogram.

UDC Universal Decimal Classification. An expansion of the Dewey decimal classification system (DDS,2). It is sometimes known as the Brussels system.

UDF Unkinkable Domestic Flex. A particular type of flexible cable which is rubber sheathed and cotton covered. It is suitable for hand appliances subject to low mechanical stress in dry situations on domestic premises.

UDK User-Defined Key. A key on a computer keyboard (KB,1) that remembers and stores a series of keystrokes, allowing the user to save the keystrokes needed to perform a specific operation and then initiate them in the proper sequence by pressing only one key.

UDLC Universal Data Link Control. A synchronous protocol, developed by Sperry Univac in the US.

UF Ultra Filtration. The technology for molecular/colloidal separations. It employs a skinned membrane structure that has semipermeable properties.

UHF Ultra-High Frequency. A frequency occurring between 300 and 3000 MHz and forming part of a larger continuous range of frequencies. UHFs are considered by international agreement to be radio frequencies in band number 9 and have wavelengths of between 10 cm and 1 metre.

UIT Union Internale des Télécommunications. This organization was originally founded in 1865 as the Internal Telegraph Union, which later became the International Telecommunication Union (ITU). In 1947 it became a special agency of the United Nations. The membership of the UIT at present numbers over 150 countries and its headquarters are in Geneva, Switzerland.

UJT UniJunction Transistor. A three-terminal device consisting of an n-type bar (*see* n-type,1) terminated with two connections B and B (*see Figure U.1*). Approximately halfway along there is a diffused p-type (*see* p-type) region which acts as the transmitter. UJTs are widely used for oscillator circuits in silicon-controlled rectifiers (SCR,1) firing circuits.

UKITO United Kingdom Information Technology Organization. A collection of UK companies that are concerned with the applications of information technology (IT).

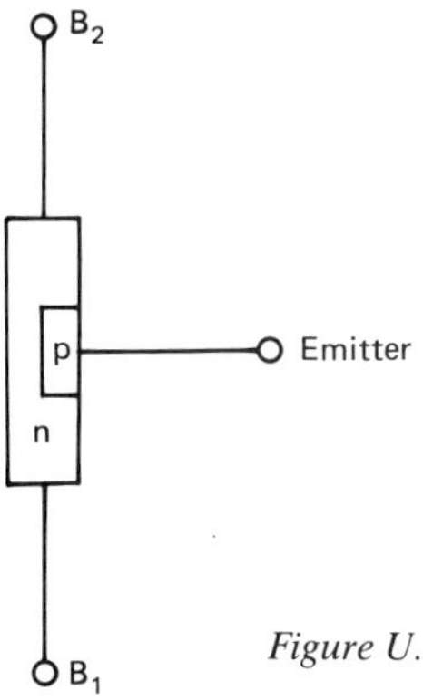

Figure U.1 Construction of a unijunction transistor

UL Underwriters Laboratories. An organization in the USA that is primarily concerned with qualifying equipment to their rigorous safety standards.

ULA Uncommitted Logic Area. Consisting of partially processed wafers of standard chip (CHIP) design, the ULA provides a compromise between the standard and custom device, as the customer is able to specify the interconnections that are necessary to provide the particular functions required.

ULSI Ultra-Large-Scale Integration. A degree of magnitude greater than very-large-scale integration (VLSI). Where VLSI microprocessors (MP) contain tens of thousands of components, ULSI microprocessors contain in the region of 100 000 components.

UMF Ultra MicroFiche. Pages of text and graphics photographically reduced with such small images that 3000 pages can be mounted on a 4 × 6 inch film.

UMI Ultra MIcrofiche. *See* UMF.

UMOS U-channel Metal-Oxide Semiconductor. So called because of the structure of its channel, the UMOS has now been abandoned in favour of the double-diffused metal-oxide semiconductor (DMOS) structure, due to what can be a difficult and critical production process.

UNIX Developed by Bell Laboratories, this particular collection of procedures for the operation of a computer is general purpose and has been designed to be multi-user.

UOSAT University Of Surrey SATellite. A communications satellite which transmits information concerning solar radiation in English, from an on-board speech synthesizer.

UPC Universal Product Code. A symbol comprising a series of parallel lines which represent a 10 number code, that has been adopted as standard by the food industry and can now be found on most food items. The first five numbers in the code indicate the item's manufacturer, the second five identify various features including size and brand, although the price is not included in the code. The code is read by an optical scanner which senses light reflected from it.

UPS (1) Uninterrupted (or Uninterruptible) Power Supply. A system designed to provide continuous power (normally of computer quality) in all abnormal power situations. UPS problems can cover millisecond (ms) fast power dropout to long-term blackouts lasting hours or days but generally less than an hour's duration; rapid dips in voltage (V,2) to long-term lows; and quick spikes in power to long-term high-voltage levels.

(2) UPper Side band. The side band containing all the frequencies above that of the amplitude-modulated (AM) carrier.

UPT Universal Portable Telephone. The UPT is envisaged as part of a new family of cordless telephones (CT) having several layers of sophistication. In its widest application the UPT is required to interface anywhere in the world to suitable telepoints.

UPU Universal Postal Union. An agency of the United Nations that promotes international co-operation between the postal services of different nations.

URL User Requirements Language. A high-level computer programming language that permits the user to employ a notation with which he is familiar.

US Unit Separator. A control character employed to designate the logical segregation of, or to indicate the boundaries of, items of data or information.

USAS United States of America Standards Association. The former name of the American Standards Association (ASA).

USASCII United States of America Standard Code for Information Interchange. Usually known as ASCII (*see* ASCII).

USASI United States of America Standards Institute. The former name of the American National Standards Institute (ANSI).

USART Universal Synchronous/Asynchronous Receiver/Transmitter. A peripheral device chip (CHIP) used for data communications. Programmed by the central processor unit (CPU) it can operate using almost any serial data technique. Accepting data characters from the central processor unit in parallel format, it converts them into a continuous serial data stream for transmission, whilst simultaneously being able to receive serial data streams and convert them into parallel data characters for the central processor unit.

USITA United States Independent Telephone Association. Consisting of independent telephone companies from a number of countries, this US-based organization deals with regulations and technical standards.

USRT Universal Synchronous Receiver/Transmitter. A general purpose communications interface that allows a high-speed synchronous communications device to transmit data to, and receive data from, a microcomputer system. It connects to the device via separate lines.

UT 200 A synchronous protocol developed by CDC.

UTE Union Technique de l'Électricité. The French national standards authority, that has a system of specifications operating in a way similar to the MIL specification (MIL,2).

UUO Unimplemented User Operation. An instruction on some computer systems issued by the user program to request the supervisor program to perform a function that the user program is not permitted to perform itself.

UUT Unit Under Test. Describes a functional printed circuit board (PCB) assembly that receives test or inspection. The designation UUT is often given to assemblies that are receiving analogue in-circuit and functional printed circuit board assembly test via automatic test equipment (ATE) systems.

UV Ultra-Violet. High-energy electromagnetic waves ranging in a wide band of wavelengths from 380 nanometres (nm) (which is just beyond the visible blue-violet) to about 10 nanometres. Radiation from the sun contains ultra-violet rays which ionize the upper atmosphere of the earth producing the ionosphere.

UVCB Under-Voltage Circuit-Breaker. A circuit-breaker that comes into operation when the supply voltage (V,2) in a particular circuit (CCT) drops below a predetermined value.

UV ERASER Ultra-Violet ERASER. A device, often in the form of a bench top cabinet, that provides a concentrated source of ultra-violet (UV) light. It is used to erase the information stored semi-permanently within UV sensitive devices such as ultra-violet erasable programmable read-only memories (UV PROM).

UVP Under-Voltage Protection. The action of defending a circuit (CCT) from voltage (V,2) lower than predetermined requirements and often associated with some form of standby power source. *See* UVCB, under-voltage circuit-breaker.

UV PROM UltraViolet erasable Programmable Read-Only Memory. A programmable read-only memory (PROM) that permits information to be stored semi-permanently, erased by concentrated shortwave ultra-violet (UV) and then reprogrammed with new information. The device is housed in a dual-in-line package (DIP) with a quartz top that is transparent to shortwave UV light.

V

V (1) Volt. Le Système International d'Unités (SI) unit of potential difference (PD), electromotive force (EMF) and electrical potential. It can be defined as the difference of potential between two points on a conductor when the current flowing is one ampere (A) and the power dissipated between the points is one watt (W,2).
(2) Voltage. The potential difference (PD) between two points in a circuit (CCT) or device.
(3) Red colour difference signal. The luminance signal employed in colour television transmission is the sum of all three colours. V is the colour difference signal (R-Y) modulated on the carrier, where R is red and Y is the luminance signal. Two such signals, U and V, are employed to derive the third.

V 24 A standard for the interconnection of equipment, similar to RS-232 (*see* RS-232). In fact they are almost identical, V24 being a Comité Consultatif International Télégraphique et Consultative Comité (CCITT) European standard and RS-232 a US standard. It specifies the way asynchronous connections should be made between computers and peripheral equipment such as printers and modultor/demodulators (MODEMs).

V35 An interface recommendation from the Comité Consultatif International Télégraphique et Téléphonique (CCITT). It employs balanced circuit (CCT) techniques for transmission of high data rates.

VA Volt-Amperes. The product of root-mean-square (RMS,2) Volts (V,1) and root-mean-square amperes (A,1).

VAB Voice Answer Back. An audio response unit that provides voice responses, from a pre-recorded digitally coded voice or disk-storage vocabulary, to telephone terminal enquiries. It can link a computer system to a telephone network.

VAC (1) VACuum. A perfect vacuum is a space from which gas has been removed and the pressure is zero. High-vacuum electron tubes achieve pressures of 10^{-6} mm Hg (1 mm Hg = 133.32 N/m^2).
(2) Voltage, Alternating Current. The potential difference (PD) between two points in a circuit or device, with a current (I) that has its direction in that circuit or device reversed with a frequency (*see* Hz, hertz) that is independent of circuit constants, i.e. an alternating current (AC,1).

VACC Value-Added Common Carrier. An organization that provides and sells the services of a value added network. These networks are built using traditional common carriers, connected to computers which permit new types of telecommunication tariffs to be offered. The network can be packet switched or message switched. The services offered include transmission of data charged for by the packet and the transmission of facsimile (FAX) documents.

VAD VApour phase epitaxial Deposition. The action of depositing materials during a vapour phase oxidation process. *See* VPE, vapour phase epitaxial.

VAL VALue. A function in beginners' all-purpose symbolic instruction code (BASIC), it converts a 'number' string into the actual number.

VAN Value-Added Network. *See* VACC, value-added common carrier.

VAR VARiable parameters. Parameters in Pascal (PASCAL) are distinguished from parameters by placing the word VAR before them at the start of a procedure.

VARACTOR VARiable reACTOR. A semiconductor diode (D,5) that behaves as a voltage dependent capacitor (C,6), when operated with reverse bias. The device can be used in low power applications, for instance varactor tuning. High power versions are employed as almost lossless frequency multipliers. For very high frequency (VHF) applications, gallium arsenide (GaAs) is used, but generally silicon is the material employed.

VARISTOR VARiable resISTOR. Usually of a semiconductor type, these resistors (R,2) have a symmetrical current (I) voltage (V,2) characteristic which can be formed by connecting two p-n junction (p-n) diodes (D,5) in parallel. This arrangement is very non-linear as its resistance (R,1) varies greatly with voltage.

VARMETER Volt-AmpeRes METER. Calibrated in reactive volt-amperes, this instrument measures the reactive component of volt-amperes in a circuit (CCT).

V band Microwave (μW) frequencies occurring between 46 and 56 gigahertz (GHz) and forming part of a larger continuous range of frequencies covering 0.225 to 100 gigahertz, which are usually subdivided into bands designated by letters. V band frequencies have wavelengths of between 0.652 and 0.536 cm.

VCA Voltage Controlled Amplifier. A particular circuit containing one or more active devices, that has been designed to amplify voltage (V,2) wave forms.

VCC Video Compact Cassette. Developed by Phillips, this is a particular system for recording television (TV(a)) programmes and other video information onto magnetic tape (MAGTAPE) housed in a plastic enclosure.

VCO Voltage Controlled Oscillator. Having a frequency controlled by the adjustment of an input (I/P) voltage (V,2), this oscillator forms part of a phase-locked loop (PLL) system.

VCR Video Cassette Recorder. A domestic videotape recorder using ½ inch tape cassettes. There are three similar although not necessarily compatible systems: Betamax, V2000 and Video Home Systems (VHS).

VDC Voltage, Direct Current. The potential difference (PD) between two points in a circuit or device with a current (I) that has a unidirectional flow in that circuit or device with a substantially constant magnitude and direction, i.e. a direct current (DC).

VDE Verband Deutscher Electrotechniker. A West German certified test laboratory that qualifies or approves equipment and components for commercial and home use throughout most of Europe. The agency qualifies equipment to both radio frequency (RF) emission and safety standards.

VDI (1) Visual Display Input. The information received by a visual display unit (VDU).
(2) Video Display Input. The information received by a unit used to display (D,6) video information, that is a device equipped with a cathode ray tube (CRT), normally a television (TV(c)) receiver or a monitor.

VDL Vienna Definition Language. A computer language used to define the semantics of other programming languages.

VDM Visual Display Module. A liquid crystal display (LCD) in the form of a matrix, supplied in a front panel of a unit which is tailored to suit Eurocard rack mounting equipment ranges.

VDR Voltage Dependent Resistor. Having a resistance (R,1) that decreases greatly with an applied voltage (V,1) increase, these non-linear resistors (R,2) are often constructed of silicon carbide pressed with a ceramic binder, formed into disks or rods and fired at 1200 degrees Celcius or centigrade (C,8.)

VDT Visual Display Terminal. Equipment with a cathode-ray tube for the visual display (D,6) of information, normally connected to a keyboard (KB,1) that is used for inputting information.

VDU Visual Display Unit. A device that provides a primary output method for alphanumeric or graphical information visually on the face of a cathode-ray tube (CRT), under computer program control. Data may be input by a user via a keyboard (KB,1) and/or some other manual method such as a light pen, cursor controls or function buttons.

VF Voice Frequency. Frequencies used for the commercial transmission of speech are generally accepted to lie within the range of 200 to 3000 hertz (Hz), although a VF may be described as any frequency lying within the human audible range.

VF band Voice Frequency BAND. A frequency range between 200 and 3000 hertz (Hz) that can provide quite intelligible and adequate speech transmission.

VFC Voltage-to-Frequency Converter. A device which accepts an input (I/P) as an analogue voltage (V,2) or current (I) signal and provides an output (O/P) as a train of pulses or square waves at a frequency which is proportional to the input value.

VFO Variable Frequency Oscillator. An oscillator that has an output (O,2) frequency capable of being varied quite considerably.

VGR Video Graphics Recorder. Often produced as a desk-top unit, a VGR provides hard copies of video displays (D,4). It has wide applications across computer graphics, engineering, drafting, design analysis, medical and industrial imaging.

VHD Video High Density. A video-disk system developed by JVC that employs a laser (LASER) to record signals on a disk. It is a capacitive system in which the pick-up is guided along the appropriate track by special signals.

VHF Very High Frequency. Frequencies occurring between 30 and 300 MHz and forming part of a larger continuous range of frequencies. VHFs are considered by international agreement to be radio frequencies in band number 8 and have wavelengths of between 10 and 1 metres.

VHS Video Home System. A widely used ½ inch video cassette format that was developed by the Matsushita Electronics Corporation.

VIA Versatile Interface Adaptor. A particular type of integrated circuit (IC) that enables some microcomputers to handle a variety of signal conversions between the computer and the outside world.

VIP Visual Information Protocol. A synchronous protocol developed by Honeywell in the US.

VISC Video dISC. An early video-disk system that employed a mechanical pick-up.

VITS Vertical Insertion Test Signals. *See* insertion test signals (ITS).

VLF Very Low Frequency. Frequencies occurring between 3 and 30 kHz and forming part of a larger continuous range of frequencies. VLFs are considered by international agreement to be radio frequencies in band number 4 and have wavelengths of between 10 km and 100 km.

VLP Video Long Player. A video-disk system developed by Philips/MCA.

VLSI Very-Large-Scale Integration. The design and production of integrated circuit (IC) or microcircuit chips (CHIP). The solid state circuitry is produced on a single piece of material, normally silicon. VLSI describes circuits with a capability of between 16 kilobits (KB,2) and one megabit (MB).

VM Virtual Memory. A technique that provides the virtual appearance of a larger main memory, by permitting the user access to a second disk, which is an extension of the main memory. By employing this method a larger memory is simulated than actually exists.

VME (1) Virtual-Memory Environment. A virtual-memory (VM) operating system (OS) that is available on some computers.
(2) Versa Module Europe. A bussing protocol for digital applications developed by Mostek, Motorola and Signetics. It defines the mechanical printed circuit board (PCB) dimensions, which are Eurocard based, the connectors, which are to DIN 41612 (*see* DIN), and the highways, protocol and enabling systems. It has been designed so that various manufacturers' equipment can be interfaced using a backplane system.

VMOS V-channel Metal-Oxide Semiconductor. So called because of the structure of its channel, the VMOS has now been abandoned in favour of the double-diffused metal-oxide semiconductor (DMOS) structure, due to what can be a difficult and critical production process.

VNL Via Net Loss. The net losses of trunks in the long distance switched telephone network of North America. The trunk is said to be in a VNL condition when it is an intermediate trunk in a longer switched connection.

VODAS Voice-Operated Device Anti Sing. A system that permits transmission in only one direction at a time, to prevent unwanted self-oscillation (singing) in a two-way telephone circuit.

VOGAD Voice Operated Gain Adjusting Device. A device employed in some radio systems to remove fluctuation from input (I/P) speech and to send it out at a constant level. No restoring device is needed at the receiving end. It is similar to a compandor.

VOL VOLume. The magnitude of the complex audio frequency signals in an audio frequency (AF) transmission system.

VOM Volt Ohm Milliammeter. A test instrument for measuring voltage (V,2) resistance (R,1) and current (I). It is usually portable and has an analogue meter as a readout.

VOX Voice Operated device. A circuit or device that automatically actuates a desired operation or function when in the presence of an audio signal. Often used to give automatic changeover on many transmit/receive units.

VPE (1) Vapour Phase Epitaxial. A method of producing optical fibres (OF) of low loss, based on the principle of vapour phase oxidation. Volatile gaseous feed materials are delivered to a reaction zone in an epitaxial reactor, where they are heated in the presence of oxygen (O,1). The product of this oxidation is a metal oxide soot which is subsequently fused to form a high purity glass.

(2) Vapour Phase Epitaxy. A semiconductor fabrication process of crystal growth, in which crystalline layers are grown onto a seed or substrate from chemical vapour of the semiconductor material, or from mixtures of chemical vapour containing the semiconductor material.

VRAM Video Random Access Memory. Similar in outward appearance to an ordinary 8-binary digit (BIT) random-access memory (RAM) a typical VRAM can be connected directly to the address and data bus of any bus-organized system, to provide composite video signals from microprocessor (MP) commands.

VRC (1) Vertical Redundancy Check. A parity check carried out on individual characters within a message block. It tests whether the sum of the binary digits (BITs) in a string of binary digits is odd or even.

(2) Voice Recognition Chip. When included with other components in a system, a typical voice recognition chip (VRC) can provide a vocabulary of 100 discrete words or phrases. The overall system formed is speaker dependent and is trained for the given vocabulary by the user.

VS Virtual System. A computer system that employs a method of operation such that it appears to the user to be accessing more main storage than the system actually contains. *See* VM, virtual memory.

VSB Vestigial Side Band. A system of amplitude modulation (AM) transmission in which one side band is transmitted in full, and the other side band (usually that corresponding to the lower modulating frequencies) is reduced in magnitude to a vestige. The attenuation is due to some of the spectral components of the side band, which are normally those relating to higher frequencies of the modulating wave. It is extensively used for television (TV(a)) transmission.

VSBS Very Small Business Systems. Computer systems priced at the lower end of the market. Besides the needed data-base storage and retrieval, the primary use of these small systems is in one or more of the following areas: technical and scientific problem solving, business systems, distributed data processing (DP,1) and word processing (WP).

VSO Very Small Outline. *See* small outline (SO).

VSS Voltage for Substrate and Sources. The common ground employed in metal-oxide semiconductors (MOS) circuits.

VSYNC Vertical SYNChronization. A system employed to ensure that the vertical movement of an electron beam in a cathode-ray tube (CRT) is locked to an applied signal. In television (TV(c)) it is called field synchronizing and the specific signal is employed to determine the vertical position of an image displayed.

VT (1) Vertical Tabulation. A specific control character that is employed to shift a printing or display (D,6) position a predetermined amount perpendicular to the printing or display line.

(2) Voltage Transformer. An instrument transformer (XFMR) used for the transformation of voltage (V,2), designed and operated similarly to a power transformer. It is also known as a potential transformer. Often used in power or test circuits, the device reproduces the voltage and current (I) values of a circuit.

VTAM (1) Virtual Telecommunications Access Method. Teleprocessing input/output (I/O) software that uses virtual techniques (*see* VM). These techniques can be used to produce a desired end result, but the user need not know which is employed. They can make the memory larger than the real memory, the difference being made up by the software moving information rapidly in and out, to and from a backing store.

(2) Vortex TelecommunicAtions Method. A software package used in data communications that simplifies and organizes data communications programming to serve remote workstations for a host computer.

VTOC Volume Table Of Contents. A list of the elements held on a computer disk directory.

VTR (1) Video Tape Recorder. Reel-to-reel machines, i.e. equipment that uses open reels of tape, as opposed to those that use video cassettes or cartridges. (*See* VCR.)

(2) Video Tape Recording. The action of using a video tape recorder (VTR,1).

VU Volume Unit. A measure of audio signal levels. For a sinusoidal signal one volume unit equals one decibel (dB). VU meters are calibrated in volume units, which measure the total energy of an input signal rather than the peak and are extensively used in audio tape recording equipment. Also known as 'voice unit'.

VXO Variable CRYSTAL Oscillator. A crystal controlled oscillator with an output than can be varied, but only over a limited range.

W

W (1) Tungsten. A heavy metal (atomic number 74) possessing an extremely high melting point and extensively used in the formation of filaments for bulbs. It is also widely used to form thermionic cathodes.

(2) Watt. The unit of power in the Système International d'Unités (SI) system of units, it is a rate of energy transfer, of one joule per second. It can be defined as the rate of transfer or conversion of energy when a current (I) of one ampere (A) flows between two points which have a potential difference (PD) of one volt (V,1). The watt is also considered as the rate of work, at one joule per second and therefore also has a mechanical definition.

WACK Wait before transmitting positive ACKnowledgement. Transmitted by a receiving station, this signal is used to indicate it is temporarily not capable of accepting a message.

WADS Wide Area Data Service. A data transmission service operating on a network similar to those serving a wide area telephone service (WATS) in the US.

WAN Wide Area Network. A computer network that covers a large physical area, even an entire continent. It has come into common use to make a distinction from local area network (LAN).

WARC World Administration Radio Conference. An international organization concerned mainly with the allocation of frequencies, but concentrating on those involved with satellite communications.

WATFOR WATerloo FORtran. Developed at the University of Waterloo in Canada, this is a version of formulae translator (FORTRAN) which provides a high rate of throughput.

WATS Wide Area Telephone Service. US telephone companies provide this service that enables customers to make calls to telephones in a specific zone on a dial basis for a flat monthly charge, using an access line. The WATS system divides the USA into six zones.

Wb WeBer. The unit of magnetic flux in the Système International d'Unités (SI) system of units. The flux is measurable in terms of volt/seconds, and 1 Wb = 1 Vs. *See* V,1 and s.

W band Microwave (μW) frequencies occurring between 56 and 100 gigahertz and forming part of a larger continuous range of frequencies covering 0.225 to 100 gigahertz (GHz) which are usually subdivided into bands designated by letters. W band frequencies have wavelengths of between 0.536 and 0.3 cm.

WCS Writable Control Store. Enabling a user to alter dynamically a microprogram, this control store is implemented with a random-access memory (RAM).

WDC World Data Centre. One of a number of international stations for promotion and collection of scientific information.

WDM Wavelength Division Multiplexing. A method where more than one signal is transmitted on a channel (CHNL(a)) by dividing the available transmission frequency into narrower bands. A range of separate single channels are thus derived from splitting up a wider bandwidth into several narrower bandwidths. Also known as 'frequency division multiplexing' (FDM).

WEG Wind Energy Group. A consortium set up by Taylor Woodrow, GEC and British Aerospace in the UK for the development of medium to large scale wind powered electricity generating systems.

W/F Wow and Flutter. Fluctuations in pitch caused during recording and or playback of film, disc or tape caused by variations of speed in the transport systems.

WIMP Windows, Icons, Mouse and Pull down menus. The generic name given to a system designed to offer a friendly interface between the operating system (OS) and the user by employing pictures to depict files of different types and allowing them to be chosen and executed by the use of an on-screen pointer. *See* GEM.

WIP Work In Progress. Generally used to describe circuits (CCTs) or sub-assemblies in the production environment. It can refer to items passing through an item of automatic test equipment (ATE) or a piece travelling through a manufacturing process.

WISE World Information Systems Exchange. Arranged to encourage international collaboration on the exchange of data between a large number of institutions.

WM Word Mark. An indicator used to signal the beginning or end of a word.

WOOFER A low-frequency loudspeaker seldom used alone. This relatively large audio reproduction loudspeaker is designed to operate on relatively low frequencies in high-fidelity (HI-FI) sound reproduction. Often it is part of a larger loudspeaker assembly comprising two or more speakers (*see* TWEETER).

WP Word Processing. Specific operating systems that include interactive information-retrieval systems, management information systems, typesetting, text editing and translation.

WPI World Patents Index. A store of information held on computer files containing patent information from all the major nations involved with research and development (R&D).

WPM Words Per Minute. A measure of speed in telegraph systems, describing the operating speed of teletypewriters (TTY) and computer printers.

WRITELN WRITE LiNe. A statement in the Pascal (PASCAL) programming language, that has the same effect as 'WRITE', but causes a 'new line' to follow the output list.

WRU Who aRe yoU. Transmitted from a data terminal or telegraph, this signal is employed to initiate the automatic return transmission of an answer-back code from the called terminal. As this control character requests a telecommunication device to identify itself, it provides a level of security.

WS Working Storage. An area on a storage disk employed to hold dynamic or working data. It is contrasted with the reserved area that contains permanent information and the user area that contains semipermanent storage. It is also known as *working space*.

WYSIWYG What You See Is What You Get. Refers to an application of word processing (WP) which offers precise control of formatting and fonts without the need for embedded commands. Thus, precisely what appears on the screen is produced on a printer.

X

X Reactance. That component of the impedance of an alternating current (AC,1) circuit which is due to its inductance or capacitance.

X25 A recommendation from the Comité Consultatif International Télégraphic et Téléphonique (CCITT). It specifies the user interface employed with public packet switching networks.

X band Microwave (μW) frequencies occurring between 5.2 and 10.9 gigahertz (GHz) and forming part of a larger continuous range of frequencies covering 0.225 to 100 gigahertz which are usually subdivided into bands designated by letters. X-band frequencies have wavelengths between 5.77 and 2.75 cm.

X-coordinate A graphics program number which permits the computer to know the horizontal position of a particular picture element (PIXEL) it is to access on the screen of a visual display unit (VDU).

XEC EXEcute register Contents. Allowing a programmer to load a machine code into a working register, this instruction then permits the central processor unit (CPU) to treat the contents of the working register as an instruction.

XFMR TRANSForMeR. A static electromagnetic device consisting of two electric windings or circuits linked magnetically by an iron core. It uses an alternating voltage (V,2) applied to one of the windings to produce, by electromagnetic induction, a corresponding voltage in the other winding.

XLPE CROSS Linked PolyEthylene. A particular type of cable insulation, that provides improved thermal performance/ overload capacity, increased insulation resistance and lower weight per metre, when compared to other materials.

XNOR eXclusive NOT OR. A logic operation where a negative output (O,2) exists only where the input statements are different or odd. *See Figure X.1.*

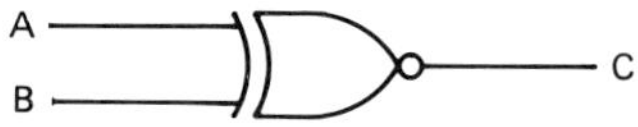

Figure X.1 The exclusive NOR (XNOR) logic symbol

X-OFF TRANSmitter OFF. (a) A signal that causes a transmitting circuit or unit to cease functioning. (b) An indicating device that signals that a transmitting device is no longer functioning.

X-ON TRANSmitter ON. (a) A signal that causes a transmitting circuit or unit to function. (b) An indicating device for signalling that a transmitting device is functioning.

X-OPEN A standard designed to encourage application portability which is supported by ten of the world's major computer companies. Based on the Unix (UNIX) operating system (OS) and an adopted definition of the C (C,1) language. Its aims are to allow users to mix and match both hardware and software between manufacturers, therefore maximizing investment.

XOR eXclusive OR. A logic operation where a true output (O,2) exists only when the input statements are different or odd. *See Figure X.2.*

Figure X.2 The exclusive OR (XOR) logic symbol

XPD CROSSPolarization Discrimination. A measure of the purity of polarization of an antenna, usually expressed in decibels (dB). It is the direction that yields the maximum (MAX) available power output (O,2) corresponding with the maximum received power, for that part of a radio system that is employed to emit radio waves and/or to extract energy from incoming radio waves.

XPI CROSSPolarization Interference. Co-channel interference that exists between a pair of orthogonally polarized signals in a radio system, that uses two signals having similar frequencies but opposite polarization.

X-position (1) A punch position, also known as the 'eleven position', this is the second row of punch positions on a punch card.
(2) *See* X-coordinate.

XPT CROSSPoinT. An electronic or mechanical switching element that can be used to extend signal and speech paths of a connection.

X-punch An X-position hole; i.e. a hole punched in the X-position (*see* X-position) of a punched card.

X-ray An electromagnetic wave that has a wavelength between 0.01 and 100×10^{-10} metres (m) with a frequency of 0.03 and 300×10^{18} hertz (Hz). The energy can penetrate directly through substances that have small or negligible attenuation but not through high-density substances such as lead and gold.

X series Transmission recommendations. Concerned with information transmission over public data networks, this is a series of recommendations made by the Comité Consultatif Internationale de Télégraphique et Téléphonique (CCITT).

XTAL Crystal. An aggregation of molecules with definite internal structure and external form of solid enclosed by symmetrically arranged plane faces. Normally a clear transparent icelike mineral. Many contemporary crystals for use in current technology are 'grown' in crystal-growing equipment.

X-Y plotter Co-ordinates plotter. Used in conjunction with a computer, this device is able to plot co-ordinate points in the form of a graph.

Y

Y (1) A star-connected three-phase power circuit.
(2) A transmitted signal that contains the brightness and detail information in colour television (TV) broadcasts.

Y-coordinate A graphics program number which permits a computer to know the vertical position of a particular picture element (PIXEL) it is to access on the visual display unit (VDU).

YIG Yitrium-Iron-Garnet. A ferrite material widely used for microwave (μW) applications. It has magnetic properties that can be altered by the amount of trace elements it contains, the most common being calcium, vanadium and bismuth. The material is used as thin magnetic film on a nonmagnetic substrate in the production of solid-state magnetic circuits such as magnetic bubble memories (MBM).

Y-position (1) A punch position, also known as the 'twelve position', this is the top row of punch positions on a punch card. (2) *See* Y-coordinate.

Y-punch A Y-position hole; i.e. a hole punched in the Y-position (*see* Y-position) of a punched card.

Z

Z Impedance. Generally described as the opposition offered by a circuit to a flow of alternating current (AC,1) in it. It is the ratio of alternating electromotive force (EMF) applied to the resultant current (I) flowing in it.

Z-clipping A computer-aided design (CAD) term. It describes the ability to specify depth parameters for a three-dimensional (3D) drawing such that all elements above or below this specified depth or depths become invisible. No alteration occurs to the database of the specific part or drawing.

ZD Zener Diode. A pn (*See* pn) diode (D,5) which exhibits Zener or avalanche breakdown at a specified value of reverse bias, and is used for voltage (V,2) reference purposes or for voltage stabilization.

ZEF Zero Extraction Force. Appertaining to connectors that require zero or minimum force to break apart mating items.

ZIF Zero Insertion Force. Appertaining to connections that require zero or a minimum of insertion force when being made. Typical units could be of the cam type of socket used for dual-in-line packages (DIP) or of the gas-tight high-pressure (GTH) design.

ZIF/ZEF Zero Insertion Force/Zero Extraction Force. Typically ZIF/ZEF connectors or sockets employ a cam so that mated connections are moved from their normal position, thus during mating or breaking apart there are no friction forces or springs to overcome. Once fully positioned, the cam is released allowing the contacts to mate. This type of connector has an inherent disadvantage in that its action does not provide a wiping action which in other connectors (*see* GTH) helps to clean the contacting or mating areas. However, this also means that there is very little if any contact wear. As a high number of cycles can be performed by ZIF/ZEF connectors, they are finding great popularity for applications in automatic test equipment (ATE).

ZIL Zigzag-In-Line. A package designed to achieve high-density installation at lower prices than existing dual-in-line (DIL) package chips (CHIP). It is achieved by bending alternate legs in opposite directions, i.e. staggering them. *See Figure Z.1.*

Figure Z.1 An example of a ZIL package showing the staggered legs (*courtesy* Mitsubishi Electric)

ZIP Zigzag-In-line Package. A package that provides high density installation at a lower price compared to existing dual-in-line package (DIP) chips (CHIP). It is acheived by staggering alternate legs of the housing.

Zn ZiNc. A metal with atomic number 30 used as an electrode in some electrolytic cells.

Zr ZiRconium. A metal with atomic number 40 that can be employed as a material because it has strong chemical affinity for other materials (a getter), in hard vacuum electron tubes.

Appendix

The British Standard for Primary Batteries, BS 397:1983, Part 1, 9.9.4, para. 3, states:

> The battery manufacturer should provide sufficient information to enable the retailer to select the correct battery for the user's application. This is specially important when supplying the first batteries for newly-purchased equipment.

Such a principle is fine – if it were adhered to – but the plethora of battery manufacturers and their products can be confusing. To help unravel any confusion, this Appendix provides a comparison of battery designations for the most popular types of battery supplied by the major producers.

In addition, the following advice concerning the care and use of batteries should help increase their life and protect equipment and its users.

Before batteries are inserted, the contacts of both equipment and batteries should be checked for cleanliness. If necessary, the contacts should be rubbed with either a clean pencil-eraser or a rough cloth, or possibly cleaned with a damp cloth and dried before being inserted.

It is extremely important that batteries be inserted correctly with regard to polarity (+ and −). One of the most common causes of dissatisfaction after replacement of a set of batteries is the reversed placement of one or more of them. In general, battery compartments are clearly and permanently marked to show the correct orientation of the batteries, and some are designed so that the batteries cannot be incorrectly orientated. Failure to comply with the instructions which are usually supplied with equipment can result in malfunction of and damage to the equipment.

It is not good practice either to use or to leave equipment exposed to extreme conditions, e.g. immediately on radiators, in a car parked in the sun, etc.

It is advantageous to remove batteries immediately from equipment which has ceased to function satisfactorily, or when a long period of disuse is anticipated, e.g. cine-camera, photo-flash, etc. Batteries should be removed from equipment also when mains current or other external source is being used to power it.

Although most batteries now on the market are provided with protective jackets or other means to contain leakage, a battery that has been partially or completely exhausted may be more prone to leak than is an unused one.

All batteries in a set should be replaced at the same time since newly-purchased and partially-exhausted units should not be mixed. Also, batteries of different electrochemical systems, grade and bands should not be mixed. Failure to observe these precautions may result in some batteries in a set being driven beyond their normal exhaustion point and thus increase the probability of leakage.

No attempt should be made (1) to revive exhausted batteries by heat or other means, (2) to charge primary batteries as to do so may cause leakage and/or explosion, (3) to recharge batteries unless they are marked 'reachargeable'. Incorrect recharging may cause rupture.

Test meters do not provide reliable comparison of service to be expected from good batteries of different grades and manufacture. They do, however, detect serious failure.

The storage area for batteries should be clean, cool, dry, ventilated and weatherproof. Batteries should not be allowed to become short-circuited, e.g. by storing them in direct contact with metal objects.

Finally, batteries must never be disposed of in a fire – they may EXPLODE.

Manufacturer	*Sizes*				
Berec Ever Ready	HP7	HP16	SP11 HP11	SP2 HP2	PP3
Daimon	242	243	241	240	244
Duracell Hellesens	MN1500	MN2400	MN1400	MN1300	MN1604
Hi-Watt IEC	LR6	LR03	LR14	LR20	6LR61 6LF22
Maxell	SUM3	SUM4	SUM2	SUM1	006P
Mazda	LK6	LK03	LK14	LK20	LK622
Panasonic	AM3	AM4	AM2	AM1	6AM6
Philips	LR6	LR03	LR14	LR20	6F22S
Ray-O-Vac	815	824	814	813	AC3
Saft Leclanche	K6	K03	K14	K20	K622
Toshiba	AM3	AM4	AM2	AM1	AM6
Ucar	E91	E92	E93	E95	522
Usa	AA	AAA	C	D	
Varta	4006	4003	4014 SP11	4020 SP2	4022 PP3
Vidor	HP7	HP16	HP11	HP2	
Wonder	KLR6	KLR03	KLR14	KLR20	KLR22

Specifications

	Battery size				
	AA	AAA	C	D	PP3
Nominal voltage	1.5	1.5	1.5	1.5	9.0
Volume (cm^3)	7.5	3.5	26.0	52.0	21.0
Weight (grams)	23.0	12.0	66.0	132.0	45.0
IEC Designation	LR6	LRO3	LR14	LR20	6F22

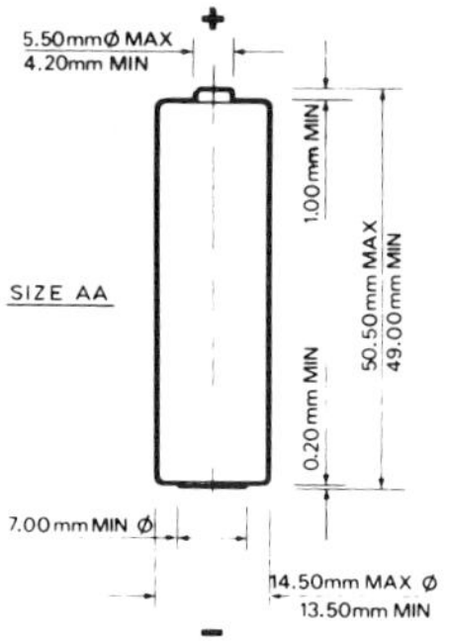

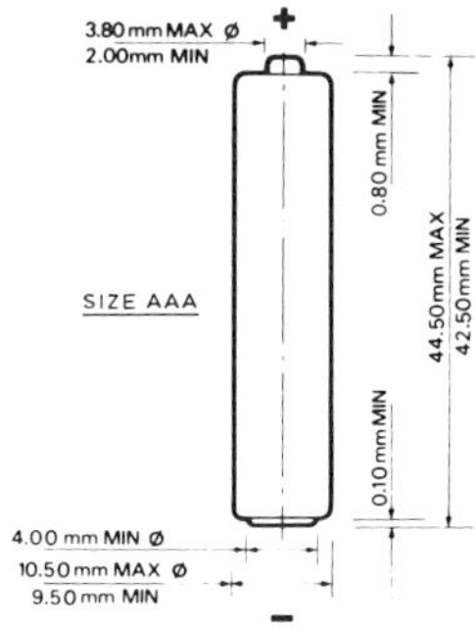

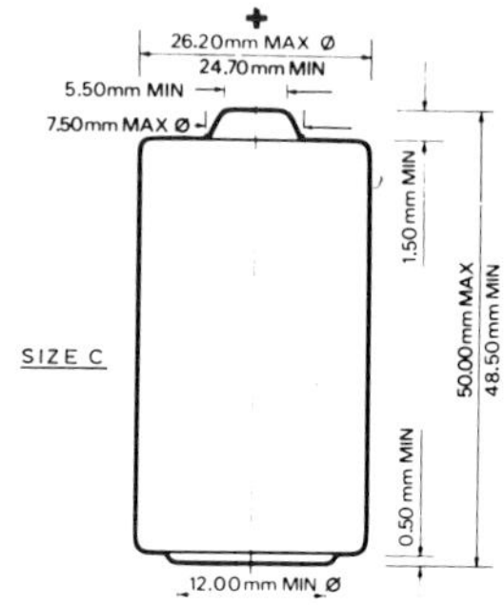

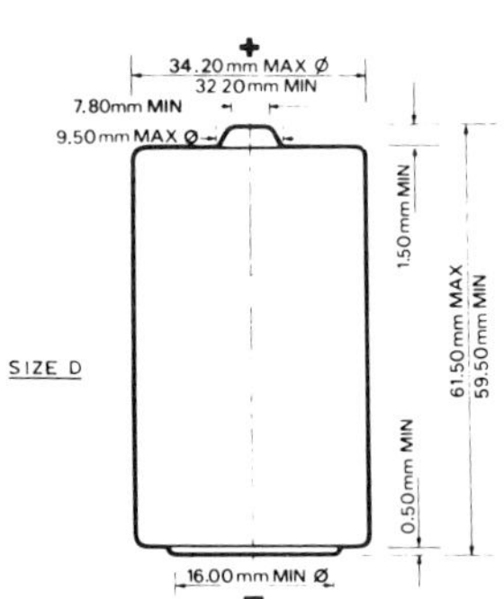

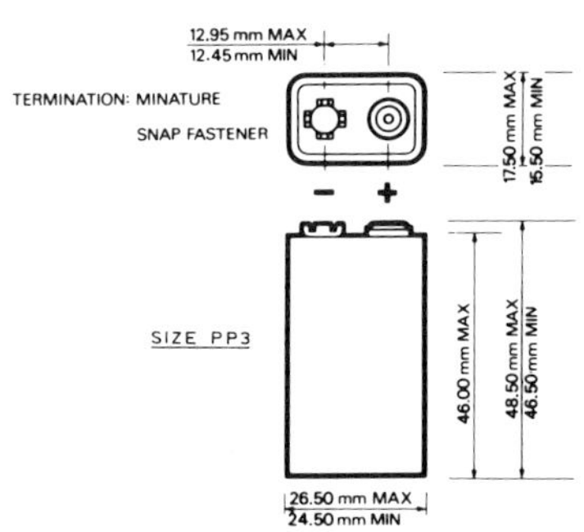